CURRENT

CURRENT

THE MAN WHO LIED TO HIS LAPTOP

Clifford Nass is the Thomas M. Storke Professor at Stanford University and director of the Communication between Humans and Interactive Media (CHIMe) Lab and the Revs Program. He is a popular designer, consultant, and keynote speaker, and is widely quoted by the media on issues such as the impact of multitasking on young minds. He lives in Silicon Valley.

Corina Yen is a design researcher and writer. She studied mechanical engineering at Stanford University, where she was the editor in chief of *Ambidextrous*, Stanford University's journal of design. She lives in the Bay Area.

The Man Who Lied to His Laptop

WHAT WE CAN LEARN ABOUT OURSELVES FROM OUR MACHINES

Clifford Nass
with Corina Yen

CURRENT
Published by the Penguin Group
Penguin Group (USA) Inc., 375 Hudson Street,
New York, New York 10014, USA
Penguin Group (Canada), 90 Eglinton Avenue East, Suite 700, Toronto,
Ontario M4P 2Y3, Canada (a division of Pearson Penguin Canada Inc.)
Penguin Books Ltd., 80 Strand, London WC2R 0RL, England
Penguin Ireland, 25 St. Stephen's Green, Dublin 2,
Ireland (a division of Penguin Books Ltd.)
Penguin Group (Australia), 250 Camberwell Road, Camberwell, Victoria 3124,
Australia (a division of Pearson Australia Group Pty. Ltd.)
Penguin Books India Pvt. Ltd., 11 Community Centre, Panchsheel Park,
New Delhi - 110 017, India
Penguin Group (NZ), 67 Apollo Drive, Rosedale, Auckland 0632,
New Zealand (a division of Pearson New Zealand Ltd.)
Penguin Books (South Africa) (Pty.) Ltd., 24 Sturdee Avenue,
Rosebank, Johannesburg 2196, South Africa

Penguin Books Ltd., Registered Offices:
80 Strand, London WC2R 0RL, England

First published in the United States of America by Current, a member of Penguin Group (USA) Inc. 2010
This paperback edition with a new preface published 2012

10 9 8 7 6 5 4 3 2 1

Two graphs and two Eye Heart and Sheep drawings by Sebastian Yen

Bush and Kerry image by Nicholas Yee

THE LIBRARY OF CONGRESS HAS CATALOGED THE HARDCOVER EDITION AS FOLLOWS:
Nass, Clifford Ivar
The man who lied to his laptop : what machines teach us about human relationships / Clifford Nass with Corina Yen.
p. cm.
Includes bibliographical references and index.
ISBN 978-1-61723-001-1 (h.c.)
ISBN 978-1-61723-004-2 (pbk.)
1. Interpersonal relations—Research—Data processing. 2. Human-computer interaction. I. Yen, Corina. II. Title.
HM1106.N38 2010
302.23'1—dc22 2010015427

Printed in the United States of America
Set in Electra Lt Std with Lucida Sans • Designed by Sabrina Bowers

To Florence Nass, Jules Nass, and Matthew Nass, and for all the family and friends who have passed away during my work on this book.

—Clifford Nass

For my loving parents, David and Julie, and my dear siblings, Jacqueline and Sebastian.

—Corina Yen

Contents

Preface

As I walked through the dormitory at Stanford University where I am a "dorm dad," I noticed Rachel, one of my freshmen, looking distraught as she knocked on her friend Jane's door. Jane opened the door and Rachel rushed in saying, "Something really horrible just happened. I desperately need your advice." As Rachel took a seat on the bed, Jane sat down in front of her computer, her desktop crowded with e-mail, multiple chat windows, Facebook, and a couple of other Web sites. The music that was playing continued to blare as Jane looked over her shoulder at Rachel and said, "It'll be okay. Tell me all about it." She then turned back to her screen and continued typing and clicking away.

I was puzzled and concerned by what I had seen, both as a dorm dad and as a researcher who studies social relationships and technology. The next day, I checked in on Rachel and asked how she was doing. "Oh, Jane was really helpful," she said with a smile. "Everything is fine now." Relieved but still confused, I told her that I had been worried because in my day, when someone was upset and came looking for advice, the first thing we would do is shut off the music, close our books, and sit down face-to-face while we talked. "That's how we showed we cared and were paying attention," I explained. Rachel gave a little laugh as she replied, "That's silly. People don't have to look at each other to know if they care. And of course we can do a lot of things at the same time. Everybody multitasks."

I still wasn't satisfied, so when I saw Jane later that day I told her about my conversation with Rachel and asked for her take on it. Jane thought nothing of her behavior: "Doing things in person is really overrated," she said. "I chat, text, and use Facebook to help my friends all the time. It's a lot more efficient, and then I can help a bunch of people at the same time. In fact, sometimes I support people I've never even seen before by giving advice on my blog when they post questions."

In this day and age, we constantly confront new social interactions amid a swiftly evolving technological landscape. It's become common to multitask during emotionally fraught situations and to give someone an online "poke" instead of a real-life hug. Facebook is the third largest "country" in the world. In just one week, more than a hundred million people will have played one online game or another using a "secret identity," often with a different gender, race, and even species than their own. While membership in churches, political parties, and civic groups has declined dramatically over the past few years, the online world is bustling with groups of "friends" and people "following" each other. Answering e-mail has developed into a 24/7 job, and now that most airlines have Wi-Fi, colleagues expect you to *always* be reachable, even while hurtling through the air at 500 mph. Texts (more than 3,000 per month are sent by the average teenager), tweets, blog posts, and e-mails have all grown explosively. The primary form of communication that is in decline? Face-to-face conversation.

As advances in technology continue to influence our day-to-day lives, it seems that the complexity of social relationships will only continue to grow, and not just in the digital space. For example, my laboratory has shown that chronic multitasking dramatically affects how people interact both online and offline. Frequent multitaskers find it very difficult to focus, even when they are pried away from technology. As a result, in a face-to-face conversation, they fail to notice emotional signals in people's voices, faces, and posture, and will often ask speakers to repeat themselves. Banning laptops in meetings can help attendees to pay attention to some extent, but chronic multitaskers—who have weak short-term memory—will nonetheless find it hard to follow the discussion.

Every change in technology seems to both initiate a new array of social rules and expectations as well as affect people in ways they don't consciously notice. It's hard enough to keep up with the latest social network, iPhone application, and viral YouTube video, but how can we also stay on top of their effects on the norms and values of how to communicate, relate, and collaborate with each other in the digital world? The popular media portrays social adeptness as so simple that the only people who struggle are the nerds of *The Big Bang Theory*, but if none of us can count on our basic instincts to get us through, we all may begin to feel as socially inept as the savants we laugh at.

When scientists attempt to understand a phenomenon that changes rapidly, they look for an underlying principle that remains steady and valid despite other fluctuating variables. Accordingly, my impulse as a researcher was to ask: On what fundamental and unchanging aspect of people can we base our understanding of social relationships? What about humans is highly stable and remains unaffected by the whims of the technological and social environment? In the face of *revolutions*, what moves at the extremely slow pace of *evolution*? When I thought about it from this perspective, the answer immediately came to me: the human brain!

Although "people are like snowflakes"—every one is unique—and the human brain is plastic, when it comes to social rules, all human brains act with remarkable similarity, shaped over tens of thousands of years of human evolution. Having evolved in a profoundly *social* world, people are hard-wired with a very broad and deep set of social rules and expectations that guide our thoughts and behaviors toward and in response to each other. Neither technology nor culture can overturn this hard-wiring. While you often hear about the "twenty-first century brain," people's most fundamental social expectations and behaviors have not and cannot change even from one millenium to the next: these well-established structures run too deep.

On the surface, each person looks and acts uniquely, just as on the surface, how people communicate and interact continues to change dramatically. However, more often than not, at the level of basic human instincts and reactions, people remain the same. By focusing on these ubiquitous and unchanging characteristics, *The Man Who Lied to His Laptop* can provide strategies for finding success and happiness and making friends and influencing people that apply as much today as they will tomorrow. Because these rules are rooted in human biology, they are *always* effective, regardless of time, place, or technology.

To uncover the social rules that have been shaped by almost a hundred thousand years of human evolution, we ironically turn to a very recent invention, the one that normally gets blamed for increasing complexity and confusion: the computer. As we explain in the Introduction, through a series of more than fifty experiments in which people interact with computers, we strip away the technological and cultural artifice that disguise basic social truths about how people work, play, and connect with each other.

The chapters that follow will describe the most fundamental truths about relationships, ones that are richly elaborated in the human brain and not affected by technology. For example, our brains react to criticism delivered through e-mail in precisely the way our brains react to face-to-face criticism. Research demonstrates there are only a few basic personality types, and identifying them is just as important when you work with a colleague across a conference table as when you fight an avatar in *World of Warcraft*. People can use the same two principles to effectively build teams whether the members sit in the same office or are scattered remotely throughout the world. Through a discussion of the underlying science and detailed descriptions of how to apply that science, you will better understand the roles that personality and emotion play in your relationships, how to give praise and criticism, how to persuade, and how to build effective teams and relationships.

As the saying goes, "the more things change, the more they stay the same." *The Man Who Lied to His Laptop* shows that the more that technology advances, the more you must focus on what does not waver: the social rules and expectations that comprise the core of human existence. By leveraging these fundamental aspects of working with other people, you can better understand and build successful relationships, no matter how the world transforms around you.

The Man Who Lied to His Laptop

Introduction

WHY I STUDY COMPUTERS TO UNCOVER SOCIAL STRATEGIES

When you work with people, you can usually tell whether things are going smoothly or are falling apart. It's much harder to figure out why things are going wrong and how to improve them. People seem too complex for you to consistently make them happier or more cooperative, or to make them see you as more intelligent and persuasive.

Over the past twenty years, I have discovered that the social world is much less complicated than it appears. In fact, interactions between people are governed by simple rules and patterns. These truths aren't vague generalities, such as advice from our grandparents ("nothing ventured, nothing gained"), pop psychologists ("follow your dreams"), or celebrities ("don't take no for an answer"). Instead, in this book I present scientifically grounded findings on how to praise and criticize, how to work with different types of people, how to form teams, how to manage emotions, and how to persuade others.

I didn't set out to discover ways to guide successful human relationships. As a professor in many departments—communication; computer science; education; science, technology, and society; sociology; and symbolic systems—and an industry consultant, I work at the intersection of social science and technology. My research at Stanford University and my collaborations with corporate teams had originally been focused on making computers and other technologies easier, more effective, and more pleasant for people to use. I didn't know that I would be thrust into the world of successful *human* relationships until I encountered three peculiar problems: an obnoxious paper clip, a suspicious auditor, and an untrustworthy navigator.

In 1998, Microsoft asked me to provide evidence that it was possible to improve one of the worst software designs in computer history: Clippy, the animated paper clip in Microsoft Office. While I have often been asked by companies to make their interfaces easier to use, I had a real challenge on my hands with Clippy. The mere mention of his name to computer users brought on levels of hatred usually reserved for jilted lovers and mortal enemies. There were "I hate Clippy" Web sites, videos, and T-shirts in numerous languages. One of the first viral videos on the Internet—well before YouTube made posting videos common—depicted a person mangling a live version of Clippy, screaming, "I hate you, you lousy paper clip!"

One might think that the hostility toward Clippy emerged because grown-ups don't like animated characters. But popular culture demonstrates that adults can indeed have rich relationships with cartoons. For many years, licensing for the animated California Raisins (originally developed as an advertising gimmick by the California Raisin Advisory Board) yielded higher revenues than the actual raisin industry. The campaign's success in fact helped motivate Microsoft to deploy Clippy in the first place. (Bill Gates envisioned a future of Clippy mugs, T-shirts, and other merchandise.) Similarly, Homer Simpson, Fred Flintstone, and Bugs Bunny all have name recognition and star power equivalent to the most famous human celebrities. What about Clippy, then, aroused such animosity in people?

Around this same time, my second mystery appeared. A market-analysis firm asked me to explain why employees at some companies had started reporting dramatic increases in the approval ratings of *all* the software applications they were using.

I started my investigation by comparing the newly satisfied users with those who had experienced no change in satisfaction. Strangely, I found that the people in the satisfied and dissatisfied companies were relatively uniform with regard to their industries (banking versus retail), the types of computers being used (PCs versus Macs), the categories of software they worked with (programming versus word processing), and the technical skill levels of their employees (novice versus expert).

I then looked at how the researchers surveyed the companies (how often, by whom, how many times). The *only* difference I found was that the companies that had started reporting higher approval ratings had changed their procedure for obtaining the evaluation. Formerly, all of the companies had people evaluate software on a separate "evaluation" computer. Later, some companies changed that procedure and had their employees evaluate the software on the same computer they normally worked with. Those companies subsequently reported higher approval ratings. Why would people give software higher ratings on one computer as compared to another identical computer?

My third problem concerned the navigation system BMW used in its Five Series car in Germany. BMW represents the pinnacle of German engineering excellence, and at the time its navigation system was arguably well ahead of other companies in terms of accuracy and functionality. Despite that fact, BMW was forced to recall the product. What was the problem? It turns out that the system had a female voice, and male German drivers refused to take directions from a woman! The service desk received numerous calls from agitated German men that went something like this:

CUSTOMER: I can't use my navigation system.

OPERATOR: I'm very sorry about that, sir. What seems to be the problem?

CUSTOMER: A woman should not be giving directions.

OPERATOR: Sir, it is not really a woman. It is only a recorded voice.

CUSTOMER: I don't trust directions from a woman.

OPERATOR: Sir, if it makes you feel better, I am certain that the engineers that built the system and the cartographers who figured out the directions were all men.

CUSTOMER: It doesn't matter. It simply doesn't work.

Something wasn't right, but the logic seemed impregnable (give or take).

How a Sock Rescued My Research

While these three dilemmas existed in vastly different products, industries, and domains, one critical insight allowed me to address all of them. My epiphany occurred while I was sitting in a hotel room, flipping through television channels. Suddenly, I saw Shari Lewis, the great puppeteer. She caught my attention for three reasons. First, instead of entertaining children, she was on C-SPAN testifying before Congress. Second, she had brought along her sock puppet Lamb Chop (not the first "puppet" to have appeared before Congress). Third, Lamb Chop was testifying in response to a congressman's question.

In her childlike "Lamb Choppy" voice (very distinct from Lewis's Bronx accent), Lamb Chop said, "Violence on television is very bad for children. It should be regulated." The representative then asked, "Do you agree with Lamb Chop, Ms. Lewis?" It took the gallery 1.6 seconds to laugh, the other congressmen 3.5 seconds to laugh, and the congressman who asked the question an excruciating 7.4 seconds to realize the foolishness of his question.

The exchange, while leaving me concerned for the fate of democracy, also struck me as very natural: here was someone with a face and a voice, and here was someone else—albeit a sock—with its own face and voice. Why shouldn't they be asked for their opinions individually? Perhaps the seemingly absolute line between how we perceive and treat other people and how we perceive and treat things such as puppets was fuzzier than commonly believed.

I had seen that, given the slightest encouragement, people will treat a sock like a person—in socially appropriate ways. I decided to apply this understanding to unraveling the seemingly illogical behaviors toward technology that I had previously observed. I started with the despised Clippy. If you think about people's interaction with Clippy as a social relationship, how would you assess Clippy's behavior? Abysmal, that's how. He is utterly clueless and oblivious to the appropriate ways to treat people. Every time a user typed "Dear . . . ," Clippy would dutifully propose, "I see you are writing a letter. Would you like some help?"—no matter how many times the user had rejected this offer in the past. Clippy would give unhelpful answers to questions, and when the user rephrased the question, Clippy would give the same unhelpful answers again. No matter how long users worked with Clippy, he never learned their names or preferences. Indeed, Clippy made it clear that he was not at all interested in getting to know them. If you think of Clippy as a person, *of course* he would evoke hatred and scorn.

To stop Clippy's annoying habits or to have him learn about his users would have required advanced artificial-intelligence technology, resulting in a great deal of design and development time. To show Microsoft how a small change could make him popular, I needed an easier solution. I searched through the social science literature to find simple tactics that unpopular people use to make friends.

The most powerful strategy I found was to create a scapegoat. I therefore designed a new version of Clippy. After Clippy made a suggestion or answered a question, he would ask, "Was that helpful?" and then present buttons for "yes" and "no." If the user clicked "no," Clippy would say, "That gets me really angry! Let's tell Microsoft how bad their help system is." He would then pop up an e-mail to be sent to "Manager, Microsoft Support," with the subject, "Your help system needs work!" After giving the user a couple of minutes to type a complaint, Clippy would say, "C'mon! You can be tougher than that. Let 'em have it!"

We showed this system to twenty-five computer users, and the results were unanimous: people fell in love with the new Clippy! A longstanding business user of Microsoft Office exclaimed, "Clippy is awesome!" An avowed Clippy hater said, "He's so supportive!" And a user who despised "eye candy" in software said, "I wish all software was like this!" Virtually all of the users lauded Clippy 2.0 as a marvelous innovation.

Without any fundamental change in the software, the right social strategy rescued Clippy from the list of Most Hated Software of All Time; creating a scapegoat bonded Clippy and the user against a common enemy. Unfortunately, that enemy was Microsoft, and while impressed with our ability to make Clippy lovable, the company did not pursue our approach. When Microsoft retired Clippy in 2007, it invited people to shoot staples at him before his final burial.

Did the social approach also help explain users' puzzling enthusiasm for their software when they gave feedback to the computer they had just worked with? Think about this as a social situation with a person rather than with a computer being evaluated. If you had just worked with someone and the person asked, "How did I do?" the polite thing to do would be to exaggerate the positive and downplay the negative. Meanwhile, if someone else asked you how that person did, you would be more honest. Similarly, the higher ratings of the software when it was evaluated on the same computer could have been due to users' desire to be *polite* to the computer and their perception of the second computer as a neutral party. Did users feel a social pull when evaluating the computer they had worked with, hiding their true feelings and saying nicer things in order to avoid "hurting the computer's feelings"?

To answer this question, I designed a study to re-create the typical scenarios in companies that evaluate their software. I had people work with a piece of software for thirty minutes and then asked them a series of questions concerning their feelings about the software, such as, "How likely would you be to buy this software?" and "How much did you enjoy using this software?" One group of users answered the questions on the computer they worked with; another group answered the questions on a separate but identical computer across the room.

In a result that still surprises me fifteen years later, users entered more positive responses on the computer that asked about itself than they did on the separate, "objective" computer. People gave different answers because they unconsciously felt that they had to be polite to the computer they were evaluating! When we questioned them after the experiment, every one of the participants insisted that she or he would *never* bother being polite to a computer.

What about BMW's problem with its "female" navigation system? Could stereotypes be so powerful that people would apply them to technology even though notions of "male" and "female" are clearly irrelevant? I performed an experiment where we invited forty people to come to my laboratory to work with a computer to learn about two topics: love and relationships, a stereotypically female subject, and physics, a stereotypically male subject. Half of the participants heard a recorded female voice; the other half heard a recorded male voice. After being tutored by the computer for about twenty minutes, we gave the participants a computer-based questionnaire (on a different computer, of course!) that asked how they felt about the tutoring with respect to the two topics.

Although every aspect of the interaction was identical except for the voice, participants who heard the female voice reported that the computer taught "love and relationships" more effectively, while participants with the male-voiced computer reported that it more effectively taught "technical subjects." Male and female participants alike stereotyped the "gendered" computers. When we asked participants afterward whether the apparent gender of the voice made a difference, they uniformly said that it would be ludicrous to assign a gender to a computer. Furthermore, every participant denied harboring any gender stereotypes at all!

People's tendencies with regard to scapegoating, politeness, and gender stereotypes are just a few of the social behaviors that appear in full force when people interact with technology. Hundreds of results from my laboratory, as summarized in two books (*The Media Equation* and *Wired for Speech*) and more than a hundred papers, show that people treat computers as if they were real people. These discoveries are not simply entries for "kids say the darndest things" or "stupid human tricks." Although it might seem ludicrous, humans expect computers to act as though they were people and get annoyed when technology fails to respond in socially appropriate ways. In consulting with companies such as Microsoft, Sony, Toyota, Charles Schwab, Time Warner, Dell, Volkswagen, Nissan, Fidelity, and Philips, I have helped improve a range of interactive technologies, including computer software, Web sites, cars, and automated phone systems. Technologies have become more likable, persuasive, and compelling by ensuring that they behave the way people are supposed to behave. The language of human behaviors has entered the design vocabulary of software and hardware companies around the world.

Of course, this "Computers Are Social Actors" approach can only work if the engineers and designers know the appropriate rules. In many cases, this is not a problem: there are certain behaviors that virtually everyone knows are socially acceptable. On a banking Web site, for example, we all would agree that it is important that the site use polite and formal language, just as a bank teller would. For a humanoid robot, it doesn't take an expert to know that the robot should not turn its back on a person when either is speaking.

What can design teams do when they don't know the relevant rules? There are three common, though flawed, strategies. The simplest is to turn to adages or proverbs, collectively accepted social "truths." Unfortunately, adages frequently conflict: for example, "absence makes the heart grow fonder" and "out of sight, out of mind";

and "many hands make light work" and "too many cooks spoil the broth." Of course, each proverb could be good advice given particular people and particular contexts, but sayings don't come with an instruction manual explaining when they should be applied. Even when following a single adage, ambiguity makes applying it a challenge. For example, absence may make the heart grow fonder, but never seeing your sweetheart again probably wouldn't nourish your romance. Similarly, how many hands are "many" hands and how many cooks are "too many" cooks? This is reminiscent of the scene in *Annie Hall* in which Diane Keaton and Woody Allen both complain to their respective psychiatrists about how often they have sex. He says: "Hardly ever, maybe three times a week." She says: "Constantly! I'd say three times a week."

A second approach is to reflect on past experiences in order to learn from trial and error. Unfortunately, in design, as in life, you don't get many opportunities to err and try again (unless you are in the movie *Groundhog Day*, in which Bill Murray's character lives the same day over and over again until he gets it right). In addition to lacking opportunities for learning, it's hard to know what lesson to learn. For example, my first dating experience lasted three dates before the girl broke it off. I decided to learn from the experience by thinking through everything that had happened during our brief relationship.* I quickly became overwhelmed; I had made all kinds of decisions in that time, and I couldn't tell which were effective and which weren't. I deliberated for a while before coming up with the perfect solution. "Since you've dated before and I haven't," I said to her, "I'd really appreciate it if you could tell me what I did wrong so that I could learn from my mistakes." Her expression mingled pity and disgust.

Last, people try to learn by example. Another dating disaster taught me the deficiencies of this strategy. When I was twelve, a suave boy

* It is traditional to refer to a "friend" when describing an embarrassing situation involving oneself. I am adopting the opposite approach by taking the blame whether it was in fact me or someone else who was embarrassed.

won the most beautiful girl at my middle school by drawing the following on the sidewalk outside her home:

When she came outside, he pointed at the drawing and said, "I did this for you!" She was immediately enthralled.

I decided that I would adopt the same strategy to entrance my lady love. I drew this, replacing "U" with a "ewe" to impress her with my wordplay:

When the girl came outside and saw me and my pictures, she ran back into her house screaming. She had concluded that I either wanted to alert her to my love for sheep or to cut out the eyes and heart of one in a bizarre ritual of devotion.

Imitating a charismatic person is difficult—even if you don't try to "innovate" as I did—and it usually comes across as a pathetic attempt at mimicry. For example, when a charismatic person asks a series of questions about someone, it feels like sincere interest; when others do it, it can seem like stalking. Similarly, rigid imitation can become self-parody, as when one attempts to frequently use a person's first name: "Hi, Cliff. It's wonderful to have you visiting us, Cliff. Cliff, let me show you where everything is."

If you try to avoid the pitfalls of imitation by directly asking people for the secrets to their success, you run into the problem that people frequently don't know what makes them successful. For example, when one of the greatest chess masters of all time, José Raúl Capablanca, was asked why he was such a poor chess teacher even though his own play was impeccable, he answered: "I only see one move ahead . . . the right one."

Although adages, learning from mistakes, and imitating others have their limitations, there is one foolproof method for discovering rigorous and effective social rules: science. Just as the *Guinness Book of World Records* or a Google search resolves sports debates, you can resolve social rule debates by turning to the relevant psychological, sociological, communication, or anthropological findings. For example, I was working with a design team to make an SAT tutoring system. We were trying to decide whether the teaching portions of the software should appear as a one-on-one session with a personal tutor avatar or as a classroom setting with avatars not only for the teacher but for the other students.

Some designers said that a solo tutor would encourage students to pay more attention and learn more. Others argued that being part of a class might make students feel less pressured because they would be just "another student" in the class and not the sole focus of the teacher. So I turned to the social science literature on how the presence of other people affects learning. As established in the classic paper on "social facilitation" by Robert Zajonc and much subsequent research, the effect of other students depends on how confident the student is. When you feel confident, having other people present improves how well you learn and perform. However, when you feel insecure, having other people around makes you nervous and pressured so you don't learn as well. As a result, we decided to have the teaching environment be a virtual classroom but with a variable number of students.

When users were doing well on the practice tests, more students would appear at the desks, but when their practice test scores were low, there would be fewer students and more empty desks.

Because new technologies appear constantly and social science rules are numerous and difficult to nail down, I was kept busy for a number of years. As a researcher, I was the expert on the "Computers Are Social Actors" paradigm, formalizing social rules and making sure that they worked with interactive technologies. Happily, they virtually always did. I became well versed in the social science literature, uncovering more and more findings that I could "steal" and apply to computers. I often joked that I had the easiest job in the world: to make a discovery, I would find any conclusion by a social science researcher and change the sentence "People will do X when interacting with other people" to "People will do X when interacting with a computer." I constantly challenged myself to uncover ever more unlikely social rules that applied to technology in defiance of all common sense. As Bill Gates described it, "Clifford Nass . . . showed us some amazing things."

While I thought that research and consulting based on this "Computers Are Social Actors" paradigm would keep me excited and challenged for the rest of my career, eventually I became dissatisfied. I had become a researcher because I wanted to discover new things, not simply "borrow" and apply what others already knew. Furthermore, I had gotten very good at doing things I had become less interested in. Ironically, it was a seemingly trivial computer application that pushed me in a new direction.

I was working with a software company on improving its spell checker. Before the development of automatic spell correction, users would check their spelling after their document was complete. Thinking about it from a social perspective, as the spell checker went through the document, all it would ever say is "wrong! wrong! wrong!" Even when you were right—for example, when you typed in a proper name or used a word that wasn't in the spell checker's dictionary—it would say that you were wrong. And what did the spell checker do when *it* was wrong? It would simply ask you to "add the word to the dictionary" without even an apology. It was not surprising, then, that few pieces of software (other than Clippy, perhaps) created greater frustration.

So I brought together the usual cast of characters (programmers, designers, marketers, and so on) to resolve the problem. As we discussed how to improve the interface, I thought about the differences between a disparaging critic and an encouraging teacher. I felt that what users needed was a "kinder and gentler" spell checker. So I suggested that in addition to signaling errors, the system could commend users on difficult words that they had spelled correctly. For example, when it saw the word "onomatopoeia," it could say, "Wow, that's a really hard word to spell right!" "After all," I argued, "it's always nice to hear some praise."

"That's ridiculous!" one of the software engineers exclaimed. "Computers are supposed to get to the point. I don't want my time wasted hearing about everything I do correctly. In fact," she added in a scathing tone, "if you really think that's a good idea, why doesn't the computer go all the way: tell users that their spelling is improving, even if it's actually lousy?"

While the engineer thought she was making a sarcastic recommendation, what our lead designer heard was a brilliant insight. "That's fantastic!" he said. "Everyone loves a little flattery, and what's the harm? It will make people feel better about checking their spelling. Users might even try harder to spell things right in order to get more praise!"

"Just what I always wanted," the engineer replied. "An ass-kissing, brownnosing, bootlicking computer! Why the heck would I want a computer to falsely inflate my ego?"

Before they could grow even more polarized, I had the other team members chime in about what they thought about flattery. Do people like flatterers? Do flatterers seem insincere or insightful? Is flattery ignored or appreciated? As our initial conversation suggested, we found little agreement, so I decided to look at what the social science literature had to say.

When I searched, however, I couldn't find anything close to a clear answer. There were isolated mentions of sincerity, kindness, honesty, and politeness in the social science literature, but nothing that tackled the question of flattery head-on. I decided to tap into my network of social science researchers to see if someone would conduct a study on flattery for me.

Although I was friendly with literally hundreds of social scientists around the world, I couldn't find one person that would take on the research. When I asked them to explain their reluctance, most researchers told me that there was simply no way to properly study flattery. For an experiment to be clean and compelling, the researcher must keep everything else constant except the characteristic that she or he wants to study. In the case of flattery, the trickiest thing to keep constant is what people say and how they say it; after all, when two people communicate with each other, almost anything can happen! Thus, when experimenters want to ensure that each participant who comes into the lab has the same experience, they hire and train a "confederate," a person whose behavior is directed by the experimenter but who is meant to appear as if she or he were just another participant in the experiment. For example, the experimenter could have the confederate and participant work together, and then the confederate could just "happen" to flatter, sincerely praise, or criticize the participant; the experimenter could then note the actual participant's reactions.

To ensure a rigorous experiment, the confederate would have to behave the exact same way every time. This can be an insurmountable challenge. Imagine how difficult it would be to say the exact same words with the exact same facial expression, tone of voice, and body language whether speaking with a very attractive person, an ugly man covered in tattoos and piercings, an obnoxious jerk, a woman who looks like your mother, or a man who reminds you of a grade-school bully. Of course, the characteristics of the confederate could also matter: flattery means something different when it comes from a smiling versus a frowning person, a woman versus a man, or someone in a lab coat versus someone in street clothes.

In the case of flattery and other questions that involve conversation and social interaction, these inconsistencies make it extremely difficult to run a rigorous study. The problem of a fully reliable confederate also plagues such questions as how to criticize (chapter 1), whether people can effectively change manifestations of their personality (chapter 2), what happens when people become teammates (chapter 3), if misery loves company (chapter 4), and when rational arguments are more or less effective than emotional arguments (chapter 5).

The other reason my social scientist colleagues would not do the research was even more frustrating. They said that questions such as the effectiveness of flattery aren't important despite how common they are in daily life. To a social scientist, "important" means addressing some fundamental question about the human brain or basic interactions among a group of humans, not helping people to have more successful relationships. It is also harder to get funding for "applied" questions than for abstract ones. For these scientists, how many people would value the information or how relevant it would be to daily life is irrelevant.

I was crushed. All I needed to make every computer user happier, more efficient, more comfortable, and more competent were answers to relatively straightforward questions about how people feel, behave, and think—the core of social science. I wasn't worried about the theorists' objections about importance because it was clear that numerous companies found my research interesting and would provide me with a great deal of money to do it; "applied" was actually a good word in many of the circles in which I traveled.

The real problem was finding a compelling confederate. I needed someone who was social but not "too" social. The confederate had to be able to carry on a constrained conversation without the participant finding it contrived. The confederate had to behave consistently in each experimental session, unaffected by who the participant was. Ideally, the confederate's demographic or other characteristics would not affect the behavior of the participant. Above all, the interaction with the confederate had to feel natural. When framed this way, it became clear to me that human confederates were simply "too human."

I am embarrassed to say how long it took me to realize that the answer to the problem was right in front of me: *computers are the perfect research confederates!* Computers, I knew, evoke a wide range of social responses similar to those elicited by people. Computers can do the same thing twenty-four hours a day, seven days a week, without deviation. They aren't influenced by subconscious responses or unintended observations about their interaction partner. Without features such as a voice or a face that mark gender, age, or other demographic characteristics, one computer is very much the same as another. Ironically, I realized that just as studying interactions between people is the best way to discover how people interact with computers, people's interactions with computers could be the best way to study how people interact with each other.

Eureka!

Experiment:

Is Flattery Useful?

My exploration of flattery, then, became the first study in which I used computers to uncover social rules to guide how both successful people and successful computers should behave. Working with my Ph.D. student B. J. Fogg (now a consulting professor at Stanford), we started by programming a computer to play a version of the game Twenty Questions. The computer "thinks" of an animal. The participant then has to ask "yes" or "no" questions to narrow down the possibilities. After ten questions, the participant guesses the animal. At that point, rather than telling participants whether they are right or wrong, the computer simply tells the users how effective or ineffective their questions have been. The computer then "thinks" of another animal and the questions and feedback continue. We designed the game this way for a few reasons: the interaction was constrained and focused (avoiding the need for artificial intelligence), the rules were simple and easy to understand, and people typically play games like it with a computer.

Having created the basic scenario, we could now study flattery. When participants showed up at our laboratory, we sat them down in front of a computer and explained how the game worked. We told one group of participants that the feedback they would receive was highly accurate and based on years of research into the science of inquiry. We told a second group of participants that while the system would eventually be used to evaluate their question-asking prowess, the software hadn't been written yet, so they would receive random comments that had nothing to do with the actual questions they asked. The participants in this condition, because we told them that the computer's comments were intrinsically meaningless, would have every reason to simply ignore what the computer said. A third control group did not receive any feedback; they were just asked to move on to the next animal after asking ten questions.

The computer gave both sets of users who received feedback identical, glowing praise throughout the experiment. People's answers were "ingenious," "highly insightful," "clever," and so on; every round generated another positive comment. The sole difference between the two groups was that the first group of participants thought that they were receiving accurate *praise*, while the second group thought they were receiving *flattery*, with no connection to their actual performance. After participants went through the experiment, we asked them a number of questions about how much they liked the computer, how they felt about their own performance and the computer's performance, and whether they enjoyed the task.

If flattery was a bad strategy, we would find a strong dislike of the flatterer computer and its performance, and flattery would not affect how well participants thought they had done. But if flattery was effective, flattered participants would think that they had done very well and would have had a great time; they would also think well of the flatterer computer.

➤ Results and Implications

Participants reported that they liked the flatterer computer (which gave random positive feedback) *as much* as they liked the accurate computer. Why did people like the flatterer even though it was a "brownnoser"? Because participants happily accepted the flatterer's praise: the questionnaires showed that positive feedback boosted users' perceptions of their own performance regardless of whether the feedback was (seemingly) sincere or random. Participants even considered the flatterer computer as smart as the "accurate" computer, even though we told them that the former didn't have any evaluation algorithms at all!

Did the flattered participants simply forget that the feedback was random? When asked whether they paid attention to the comments from the flatterer computer, participants uniformly responded "no." One participant was so dismissive of this idea that in addition to answering "no" to the question, he wrote a note next to it saying, "Only an idiot would be influenced by comments that had nothing to do with their real performance." Oddly, these influenced "idiots" were graduate students in computer science. Although they consciously knew that the feedback from the flatterer was meaningless, they automatically and unconsciously accepted the praise and admired the flatterer.

The results of this study suggest the following social rule: *don't hesitate to praise, even if you're not sure the praise is accurate*. Receivers of the praise will feel great and you will seem thoughtful and intelligent for noticing their marvelous qualities—whether they exist or not.

The rules and principles presented in this book have emerged from using the computer-as-confederate approach to make discoveries that previous social science approaches could never uncover. One cannot fail to see the irony here. Not only are computers associated with the most unsociable responses imaginable (e.g., "Your response is invalid. Try again"), computers are stereotypically the domain of the most socially inept people. Nonetheless, computers' "deficiencies" are what make them key to understanding social behavior and discovering successful social strategies.

The experiments that I now conduct uncover surprising and powerful social rules that apply to people (as well as to computers). Whenever a clear rule does not exist in the social science literature, I nail it down through experiments pairing people with computers. The experiments present people with the same contexts—collaboration, evaluation, learning, playing—and the same human roles or characteristics—praiser versus criticizer, male versus female voices, dominant versus submissive personalities, happy versus frowning faces. The experiments include traditional measures and metrics to assess people's behaviors—standard questionnaires for personality and liking, memory tests, physiological measures of emotion. And I formalize the conclusions in terms of actionable rules that can create and support successful human relationships as well as advance the social sciences and user experience design.

This approach forces me to be ruthlessly direct and precise in the questions I ask and try to answer. A computer follows rigid steps and uses ironclad reasoning to reach exact, objective, and universal results. Thus, computer-derived rules are unambiguous, rigorous, and straightforward—making them readily usable in daily life. Because a computer is so obviously not a social presence—lacking a face, a body, emotions, and so on—if a social rule is effective for a computer, it will be even more effective when followed by a person, regardless of the situation, time, and place. For example, while a person flattered by another person might rationalize that somehow the flatterer was being sincere, the computer was obviously and unambiguously flattering: (seemingly) making random comments. Nonetheless, participants believed they did better because of it. The effectiveness of such blatant and irrelevant flattery suggests that these results are a conservative reflection of success you can attain in daily life by flattering others.

The rules I have uncovered and describe are so basic that any person (or computer) can apply them easily, and they are so broad and effective that every person (or computer) can become more persuasive, likeable, and socially successful. And while the rules are simple, they need not be followed mechanically: each rule is presented with the relevant underlying psychology so that you know how and when to apply it effectively.

I have long enjoyed the opportunity to work with designers and engineers to improve products and services, making cars safer, educational software more engaging, mobile phones more socially supportive, robots less frightening, and Web sites better able to close the deal. Now I also confer with social scientists about the "holes" in their understanding of people. In addition to improving products, I use my rigorous experiments with computers to help people evaluate others more effectively, work more smoothly with those different than themselves, manage their own and their colleagues' frustrations, and better persuade others. Combining the theories and methods of social science and cutting-edge research with computers where social science is inadequate, the insights in *The Man Who Lied to His Laptop* will help you improve your professional and personal relationships.

The discoveries presented in this book are far-reaching. You will no longer use the "evaluation sandwich"—praise, then criticism, then praise again—after learning that it is neither helpful nor pleasant. You will identify the personalities of your customers and use that information to better persuade them. You will discover why team-building exercises don't build teams, and what to do about it. You will leverage the "laws of emotion" to defuse heated situations and rally your colleagues. You will appreciate that even unintentional or meaningless inconsistencies carry great weight. The rules that emerge from the fascinating and sometimes bizarre ways that people treat computers like people will give you the tools you've always wanted to dramatically improve your day-to-day life. I invite you to join me as I move back and forth between the world of people and the world of technology, finding life-changing insights in both.

CHAPTER 1

Praise and Criticism

One of the most stressful times in any organization is "evaluation week." Although managers give employees feedback throughout the year, the period specifically set aside for telling people what they do well and what they do poorly seems to evoke a special kind of fear and loathing. People being evaluated are not the only ones who suffer: evaluators also worry about having to label every aspect of a person as "good" or "bad."

How have companies addressed the anxieties triggered by evaluation week? By asking everyone to do more evaluations. In addition to managers' evaluations of their subordinates and teachers' evaluations of their students, individuals are increasingly asked to provide evaluations of their peers and even their superiors in a process known as 360-degree feedback (in other words, you're not safe from any angle). Mandatory assessments and the documentation of these assessments have become universal.

Sometimes even more challenging than evaluating others is the increasingly widespread requirement that people evaluate themselves. Say something too nice and you're bragging; say something too critical and you're insecure. Stick to the specifics and your impact seems small; state generalities and you're hiding your mistakes. And because no one sees you exactly as you see yourself, your honest beliefs can appear to misstate reality.

No job is more immersed in evaluation than being a professor. Whenever I teach, I must provide feedback, both positive and negative, about each student's work. People inundate me with requests to review books, papers, tenure files, and presentations. I have to write letters of recommendation for every possible type of graduate school and job, even when multiple students want recommendations for the same position. Worst of all, I sometimes have to recommend students even if I don't want to. In these cases, I have been tempted to follow the tongue-in-cheek advice of Robert J. Thornton, professor of economics at Lehigh University, who suggested the following could be used for weak candidates to protect yourself from lawsuits:

- To describe a candidate who is woefully inept: "I most enthusiastically recommend this candidate with no qualifications whatsoever."
- To describe an ex-employee who had difficulty getting along with fellow workers: "I am pleased to say that this candidate is a former colleague of mine."

What goes around comes around: I have received scores of reviews of my books and over five hundred reviews of my papers, including many that have felt unfairly negative. Every quarter, all of my students evaluate my teaching. No matter how many times I receive feedback on my work, the criticism still stings, even when it comes from a faceless freshman receiving a D in my class.

Furthermore, each year Stanford asks me to provide a detailed assessment of my own performance. On the one hand, as an employee, I am supposed to put myself in the best possible light using all of the powers of persuasion that I can muster. On the other hand, as a researcher, I have been trained to "let the data speak for itself" and avoid pushing one interpretation over another; that part of me feels that anything more than my academic résumé must be biased.

Despite how frequently I evaluate people and am evaluated myself, I have agonized over evaluations for a long time. I've been told "don't be judgmental" but also that "facts without interpretation are like seeds without soil." If I provide only praise, it sounds like hagiography (the study of saints), but each criticism I add seems to jump off the page. If I am too effusive, I sound like a cheerleader; if I am too flat, it reads like I'm hiding something. And the order problem is overwhelming: is it praise before criticism, criticism before praise, praise-criticism-praise (the evaluation sandwich), or some other arcane formula? And should I do things differently when I talk about myself as opposed to others?

As with so many other domains of life, the "common wisdom" is ambiguous and contradictory. After years of struggling with giving evaluations, riding an emotional roller coaster when receiving them, and confronting ethical dilemmas when evaluating myself, I decided to search for guidance about the best ways to evaluate and to receive evaluations by investigating the social science literature, and, when the social science literature fell short, to do the research myself. Specifically, I sought to answer the following questions:

- Can you avoid giving evaluations?
- Are praise and criticism more than opposites?
- How can you most effectively deliver praise and criticism?
- How are people's perceptions and opinions of you affected by how you evaluate others and how you evaluate yourself?

Proverbs clearly indicate that you should avoid evaluating others: "Judge not lest ye be judged," "Don't look a gift horse in the mouth," "Don't judge a book by its cover," and "People in glass houses shouldn't throw stones." Although many adages warn against evaluation, discerning between good behavior and bad is the most primitive judgment that humans make and is virtually impossible to suppress. This is because categorizing someone as good or bad is part of the possibly life-or-death decision to approach or avoid that person. Thus, the judgment of positive or negative is built into our every fiber.

Scientists have traced the drive for evaluation to the center of our brains: the thalamus. Sitting on top of the brain stem, the thalamus connects to every part of the higher-thinking areas of the brain. It makes very basic judgments about whether you have encountered someone who is extremely good or bad even before information enters your formal thought processes. For example, if someone is smiling versus shouting at you, the thalamus decodes this valence—positive versus negative—and will react before you even understand what the person is saying. Once the thalamus makes a positive-versus-negative judgment, it sends a call to the action centers of your body to prepare the muscles to approach or avoid. The thalamus then passes on its interpretation of positive or negative, along with the words being spoken, to the higher (and slower) thinking parts of the brain.

The automatic and simplistic response of the thalamus to evaluation is universal. When parents see their baby smile for the first time, they feel joy even after finding out that it was due to the baby's gas. Hearing "good job" from someone, even if she or he is unfamiliar with your work, can make you feel wonderful. On the other hand, when a two-year-old cries out, "I hate you," parents' shoulders slump and they feel terrible, even though they know that two minutes later the child will probably reverse his or her feelings. And if people see an angry glance, they become anxious, regardless of the actual source of the person's anger.

When the thalamus cannot identify the valence of the evaluation—for example, if someone speaks with a straight face—it sends the information to the more sophisticated parts of the brain before signaling the body to react. Those higher-thinking processes interpret what is being said in order to make the positive-versus-negative judgment (e.g., recognizing praise versus criticism); they then send this information back to the thalamus, which subsequently guides your reactions. Thus, if your manager walks up to your desk and unsmilingly tells you in a monotone voice that you are getting a very large bonus, you will react with happiness more slowly than if your manager had bounded up with a big grin and shared the news in an enthused voice.

As a result of this human drive to judge what one encounters as positive or negative, people spend their lives praising or criticizing almost everyone they meet. As a child, you hear "good girl" and "that's a no-no." Teaching someone to ride a bicycle, shoot a basketball, or drive a car involves constant reference to whether the learner is doing well or poorly. Discussions commonly revolve around topics such as whether you like or hate another person, whether you think someone's opinion was smart or dumb, or whether an athlete's performance warranted cheers or boos. Suggestions on what to wear, whom to date, or what job to take carry an implicit message of praise or criticism. Tone of voice, a lingering glance, and the tilt of a head can all communicate whether someone wants your behavior to continue or stop. My son, Matthew, puts it even more simply: "When someone uses your full name, you know you're in trouble."

Can People Give Neutral Evaluations?

While the positive-negative dimension is clearly ingrained in evaluations and in the way we communicate, some believe that with effort you can eliminate it. I once consulted for a company that insisted my evaluations should "avoid judgments; simply indicate the extent to which each goal was met." However, the majority of seemingly "neutral" words still tilt positive or negative. For example, did employees "achieve goals" or "perform tasks"? Did they "strive" or "struggle"? Did they "discover," "find," or "stumble upon" a solution to a problem? In fact, linguists Robert Schrauf and Julia Sanchez have shown that only 20 percent of typical words in English or Spanish have a completely neutral connotation. Another 50 percent of words have negative orientations, and the remaining 30 percent have positive orientations. And what if you scrupulously limit yourself to the 20 percent? You run into a problem called the self-serving bias: even people with low self-esteem believe that they are better than they actually are along virtually any dimension.

Think about it: The standard response to "How are you doing?" is "fine," not "average" or "neutral." Literally hundreds of studies, summarized by clinical psychologist Amy Mezulis and colleagues, have demonstrated that the vast majority of people, in almost all cultures, believe that they are smarter and more attractive than the typical person: it's not just Lake Wobegon in which all of the children are better than average! Almost everyone also believes that she or he is more likely to obtain high incomes and less likely to get diabetes, be hit by a meteor, or get divorced than the average person. Therefore, people see an attempt to remove all positive and negative remarks as a negative evaluation because they actually perceive neutral as negative.

In sum, people's brains are wired to both constantly evaluate others and to interpret every piece of feedback they receive as a judgment. You cannot avoid being judgmental, and it's futile to try to give unbiased and valence-free feedback. Knowing how attention-getting evaluations are, the question is, how do people interpret and respond to positive and negative evaluations?

People do not receive positive and negative evaluations in equal and opposite ways. Longfellow's little girl with the curl highlights this: "When she was good, she was very, very good, but when she was bad, she was horrid." That is, negative is more noticeable, consequential, and extreme in every respect as compared to positive. This "hedonic asymmetry" is a natural consequence of human evolution. Compared to the vast majority of other species, humans produce few young and have the potential for very long lives. Hence, the human brain is optimized to identify and respond to bad experiences; good news can wait. From the first instant that a stimulus hits the sense organs until the brain and body fully process and resolve the experience, negative gets virtually all of people's attention; positive is merely a bit player.

You can see evidence of the power of negative in everyday experiences. Most drivers slow down to see a car wreck; far fewer pause to admire a bucolic vista. While many complain that news organizations do not cover enough happy stories, an analysis of the history of newspapers by MacArthur Award–winning sociologist Michael Schudson found that they only became popular as a medium when they started covering negative and tawdry stories. And as we will show in the Emotion chapter, it is much easier to make people cry than laugh.

Experiment:

Twenty Questions and Insulting Answers

Just how different are negative and positive? To answer this question, I extended the Twenty Questions flattery experiment described in the Introduction in which the computer would "think" of an animal, and participants would ask "yes" or "no" questions to narrow down the possibilities.

In the flattery condition of the study, the computer would praise the participant's questions, for example, calling them "clever" and "ingenious." While we told one group of participants that the feedback was highly accurate and another group that it was random, in reality, both groups received the same comments. A third group, for comparison, received no feedback at all. We found that people accepted praise they thought meaningless just as willingly as praise they thought accurate.

To determine whether the same blind acceptance would occur with criticism—that is, whether our results would change if the feedback were negative instead of positive—we added two groups to the experiment. Both sets of these users received identical negative responses: their questions were "confused," "ineffective," "foolhardy," and so on. As before, one group was told that the evaluations were accurate; the other that they were random. In other words, we now had a group receiving (ostensibly) sincere criticism and a group receiving random criticism ("calumny"). At the end of the session, we once again asked participants a number of questions about how much they liked the computer, how they felt about their own and the computer's performance, and whether they enjoyed the task.

If people accept calumny as readily as they do flattery, we would expect that the recipients of random criticism would think that they did as badly as those who received sincere criticism. If, on the other hand, participants scrutinized the random negative remarks, false criticism wouldn't affect them. They would understand, quite correctly, that the evaluation they received had no basis in fact and thus shouldn't influence their thinking.

➤ Results and Implications

Consistent with the idea that people scrutinize criticism much more carefully than praise, I found that receiving calumny did not affect how well participants thought they did. That is, there was no difference between receiving false criticism and not receiving any evaluation at all. In comparison, the people who were supposedly criticized accurately thought they did much worse. Thus, people do not automatically accept criticism: whether criticism comes from an accurate source as opposed to being randomly dispensed makes the difference between believing and dismissing.

So while people are not suckers for calumny, they are for flattery—even from a computer. When we hear something positive about ourselves, we happily accept it. We don't worry too much about either the source or the basis for the remark. And if someone delivers an evaluation with a smile or a warm tone, it further amplifies the effect: the thalamus will provide unconscious support for feelings of (potentially unwarranted) joy.

Regardless of how accurate participants believed the evaluation to be, its valence affected their perception of the evaluator: sincere praise and flattery were equally likeable, and criticism and calumny were equally detrimental to the evaluator. That is, praisers are liked and critics are hated, right or wrong. In another example of how powerful negative remarks are, I performed a study with my Ph.D. student Laurie Mason, now a professor at Santa Clara University, which demonstrated that when a newspaper quotes person A criticizing person B, people develop negative feelings about person B, person A, *and* the newspaper! When criticizing, neither accuracy, inaccuracy, nor simply repeating someone else's negative remarks gets the critic off the hook. Similarly, praise, flattery, and repetition of others' positive remarks all benefit the praiser.

Because people do not deeply consider the praise they receive, in the long run, they will not remember the specifics of the praise—although they may recall that they were praised and that they had positive feelings about it. For example, if I asked you to remember the last few people who complimented you, you probably could remember them quite well. If I asked you the exact comments they made, though, you probably would find it much harder to recall.

People remember criticism, on the other hand, very well. If asked about the last negative messages received from a computer or a person, people will generally be able to recall the slights in significant detail. For example, when you present an employee with a list of job performance criteria and a set of markings with "excellent, very good, good, fair, and poor," the "excellents" will trigger a small reaction, the "very goods" and "goods" will get a vague smile, but every "fair" and "poor" will be analyzed, interpreted, and remembered for days and even weeks. This also plays out in politics: in a study conducted by communication professors Diana Mutz and Byron Reeves, when candidates on a supposedly real political talk show were highly critical of each other, people better remembered which candidate was on which side of each issue than when the candidates did not attack each other.

One fascinating side effect of the power of negativity is that you remember less of what is said before receiving criticism because negative remarks demand so much cognitive power that the brain cannot move the prior information into long-term memory. Known as "retroactive interference," this explains why it is often difficult to give a detailed answer when asked, "What made that person yell at you?" People frequently cannot remember what they were doing just before their computer started behaving strangely—a common problem for technical support professionals attempting to troubleshoot. (Because praise and positive events do not require significant cognitive resources, they do not cause retroactive interference.)

Immediately *after* a negative evaluation, however, the brain and body go into full alert. People have a number of consequential choices after receiving a negative remark: walk away, defend, argue, escalate, physically threaten, or plead for a solution. We immediately seek information that will help guide us in our decision. So, after a negative event, our memory is actually *improved*, an effect known as "proactive enhancement." This is why you should present information you want remembered immediately after a negative remark.

When you want to give a mix of positive and negative feedback, the order is critical. Tradition states that one should give praise first to "soften up" the person before giving her or him bad news. However, this is a poor idea: although the immediate reaction to the negative remark will be softened, in a short time retroactive interference will come into play and all that will be remembered is the negative remark. It is better to present the negative feedback first and then the positive evaluation. The criticism will bring people to attention in time to listen to your praise.

An even worse prescription than praise before criticism is the so-called "criticism sandwich": 1) specific positive comments, 2) specific negative comments, and 3) an overarching positive remark. The idea here is that by bracketing the negative remarks with positive comments, you make the criticism palatable. Unfortunately, given retroactive interference and proactive enhancement, a very different outcome occurs: the criticism blasts the first list of positive comments out of listeners' memory. They then think hard about the criticism (which will make them remember it better) and are on the alert to think even harder about what happens next. What do they then get? Positive remarks that are too general to be remembered.

It is also important to consider that receiving an equal number of positive and negative remarks feels negative overall because of hedonic asymmetry and the self-serving bias. It is far better to briefly present a few negative remarks and then provide a long list of positive remarks. This can take significant effort—it's much easier to remember negative impressions—but generating lists of positive remarks is time well spent. You should also provide as much detail as possible within the positive comments, even more than feels natural, because positive feedback is less memorable.

The previous section demonstrated dramatic differences in how people scrutinize and remember criticism and praise in general. The next question is how the language you use to praise and criticize affects the reception of your evaluation.

Experiment:

A Car-tastrophe

I had been thinking about how one might research the consequences of word choice in evaluations when I learned that a group at a Japanese car company had inadvertently designed an experiment for me.

The automobile manufacturer was concerned with the dangers associated with poor driving by truckers, taxi drivers, and other professional drivers. To address the issue, they developed a system that could detect when a person was driving poorly, via the ingenious use of sensors and artificial intelligence, and then inform the driver.

Before they installed this elaborate system in production vehicles, the company decided to test it in a car simulator. They invited me to observe and help evaluate the tests. It was the nicest simulator I had ever seen: a complete automobile surrounded by 270 degrees of floor-to-ceiling screens that immersed the driver in the environment. The simulation responded flawlessly to the gas pedal and brake. It offered impressive force-feedback controls and high-fidelity surround sound—

the driver could feel every turn and bump in the road and hear remarkably accurate noises from the car and the environment. The system measured, second by second, the performance of the driver and the car as the driver dealt with various situations on the road. For my part of the research, I had prepared a wide-ranging questionnaire that addressed how drivers felt about their driving performance and the car's responsiveness and intelligence.

➤ Results and Implications

As I observed the first driver use the system, I quickly saw the negative consequences of having one's car become a "backseat driver." During the demonstration, the participant exceeded the speed limit and made a turn a little too sharply.

"You are not driving very well," the car said. "Please be more careful."

Was the driver delighted to hear this valuable information from a highly accurate and impartial source? No. Instead, the driver became somewhat annoyed. He started to oversteer, making rapid, small adjustments to the wheel; the system reported an increase in driving speed and a decrease in driving distance from the next car.

"You are driving quite poorly now," the car announced. "It is important that you drive better."

Was the driver now appropriately chastened? No. His face contorted in anger as he started driving even faster, darting from lane to lane without signaling. He could not keep the steering wheel still, swerving back and forth from one side of the lane to the other at a frightening pace, tailgating the cars in front of him. This spiral of negative evaluation, anger, worse driving, and more negative evaluation escalated.

"You must pull over immediately!" the car said. "You are a threat to yourself and others!"

At this point the driver, literally blind with rage, smashed into another car in the simulation. He was so livid I couldn't even understand what he was saying. My questionnaire was obviously unnecessary as the lesson was clear: even stunningly accurate criticism may not be constructive.

The extreme anger of the driver provides a key insight about delivering criticism: there is clearly a wrong way to do it! The system failed to be an effective evaluator because it ignored how the human body responds to negative stimuli. As discussed earlier in the chapter, the brain judges things as good or bad precisely so that when people encounter the bad, they can quickly address it by either attacking or fleeing. The body prepares itself for action in multiple ways: heart rate increases, blood pressure and adrenaline level rise, and more oxygen is inhaled. With all these energizers, you cannot expect people to calmly accept a negative evaluation—criticism readies the body to attack with words and fists or to run away. This explains the criticized driver's behavior: adjusting the wheel (something to do), driving rapidly (fast action), and tailgating (a combination of aggressive behavior and trying to alter the situation).

Fight-or-flight responses are governed by the emotional parts of the brain. These parts can demand action without consulting the higher-order, rational areas of the brain that know the "facts" of the situation. This is why criticism will often generate seemingly irrelevant statements, ad hominem attacks, scapegoating, frantic apologies, and little valuable information. It also explains why people being interrogated have the right to remain silent and why torture very frequently produces false information.

Making Criticism Constructive, Not Destructive

Given people's volatile response to negative evaluations, how should you criticize? First, the most effective negative evaluations focus criticized people's action-oriented state toward constructive ends. In contrast, simply telling someone that his quarterly reports have been consistently late gets the person riled up and ready to do something but without guidance on what to do. This can leave the criticized person either attacking you or reacting defensively. When the car insisted that the driver was performing poorly but gave no guidance on how to improve, it encouraged the driver's downward spiral.

The better approach involves coupling criticism with suggestions for improvement, presenting the person with a clear (and constructive) way to react to the criticism. For example, in addition to criticizing your subordinate for his tardy quarterly reports, you could ask that he develop a plan for timely production, suggest that he omit the most time-consuming and least relevant parts of the document, or propose that he work more hours. Real "constructive criticism" is not simply a valid assessment of a person's work—no one initially perceives even the most accurate criticism as "constructive"—it must also guide the recipient on how to act on the evaluation.

Similarly, when you deliver criticism, go deep rather than broad. By providing specific details on one problem, you provide a clear picture to the criticized individual of the appropriate solutions. You also take advantage of the proactive enhancements to memory: the listener will better remember details given after criticism. Conversely, rattling off a long list of complaints creates paralysis because it's not clear where to begin. Also, because most criticism is usually delivered from most to least important, retroactive interference will lead the least important criticism to drive the most important out of memory.

The action orientation that results from criticism also makes *when* to criticize a critical consideration. A passing negative remark doesn't allow someone to fully react. While you or the conversation have moved on, the person's urge to act has been left unresolved. This can lead to frustration, which in turn can lead to aggression or panic. Similarly, criticizing people and then telling them to "sleep on it" can feel like a "hit and run" as you abandon them with their feelings still in turmoil. And if they spend all night thinking about the criticism rather than sleeping, they will feel even more terrible in the morning. Don't force people to listen to your criticism without giving them a chance to react to you.

On the other hand, don't ask for an immediate response after you criticize someone. Anything you hear from the recipient at that point will stem directly from the emotion centers of her or his brain. Ideally, invite a brief response and schedule a follow-up discussion, giving the recipient time to fully process your feedback.

Experiment:

Praise and Calming the Car-tastrophe

After observing the extreme behavior of the criticized drivers in the car simulator study, I suggested that we have some drivers receive praise to see how they would react.

➤ Results and Implications

The results were underwhelming. When the drivers received praise, their driving performance did not follow any strong or consistent patterns: some people drove better, others drove worse. The drivers seemed more relaxed and at ease, but none of them exhibited positive emotions that approached the intensity of the anger displayed by the negative drivers.

This study suggests that praise affects behavior far less than criticism. Looking into the existing research about the phenomenon, I found that when people receive a positive evaluation, unless it is extremely positive or highly surprising, their bodies tend to relax: there is nothing to worry about. People's heart rate slows, their blood pressure lowers, their adrenaline level goes down, and their muscles relax. A simple "thank you," a "that was nice of you," or a nod of the head feels like more than a sufficient response to praise. Hence, praise rarely spurs people to greater heights, although it can help ensure continued positive behavior.

Because praise has less impact than criticism, deliver praise in ways that make it memorable. For example, the human brain loves repetition of sound (rhyme) and meter (prosody): the former supports memory and the latter makes it easier to process the meaning more deeply. This is why odes—attempts to glorify a person or thing—almost always rhyme and have clearly marked rhythms. Create positive, esteem-boosting nicknames such as "The Closer" or "Mr. Programming" that are unique (to make them more memorable) and use them regularly (if the nicknames are clever enough they will catch on with the entire group). Love poems also leverage these strategies. Arguably one of the most famous opening lines in English literature is Elizabeth Barrett Browning's "How do I love thee? Let me count the ways." The power and memorability of the line come from the fact that "I love thee" is repeated eight more times in the fourteen-line sonnet.

A second approach for enhancing positive evaluations is surprise because it gets people to pay attention and think harder about what you just said. For example, if you compliment someone on something that he or she thinks you are unaware of, it will have a bigger effect than if you keep dishing out the same obvious compliments. Slipping these surprising references into a list of more obvious positive remarks is even more effective. This is another reason why flattery works so well: it is surprising because it might not even be true!

In addition to how and when you deliver evaluations, the orientation of the evaluated person also can affect how he or she receives your evaluation. Stanford psychologist Carol Dweck has discovered that people's fundamental views about the nature of success color how they interpret praise and criticism. Dweck calls these fundamental views "mindsets." People with a fixed mindset believe that intelligence and abilities are innate qualities, essentially carved in stone the day you are born. Extending effort to improve is a waste of time: you either have it or you don't. As a result, when people with a fixed mindset hear the evaluation, "Your performance on that activity was poor," they generally decide to avoid the activity, believing that they cannot improve. Conversely, people with a growth mindset believe that failure is changeable and success can be cultivated through one's efforts. When growth mindset people are told, "Your performance on that activity was poor," they ask themselves, "What can I do to be better in the future?" While people with a fixed mindset agree with comments such as, "You can learn new things, but you can't really change how successful you will be," people with a growth mindset believe, "No matter how much talent you have or don't have, you can always improve if you work at it."

Experiment: ------------------------------

Framing Failure as Friend or Foe

Dweck has found that a growth mindset positively affects self-confidence, affinity for learning, ability to overcome challenges, and resilience in the face of setbacks. People with a growth mindset can, as Rudyard Kipling put it, "meet with triumph and disaster and treat those two impostors just the same" because they believe that effort can lead to success and can overcome failure. As a result, all other things being equal, growth mindset employees are more likely to become valuable contributors in the long run.

Typical discussions about mindsets betray a fixed mindset. That is, once hearing about the concept, people talk about "fixed mindset people" and "growth mindset people" as if mindsets were an immutable trait. However, can the mindset of an evaluator affect the mindset of the person she or he is evaluating? To examine this possibility, I, along with Ph.D. student Shailendra Rao, designed a study to determine whether the content of an evaluation can encourage a healthy growth mindset, at least with respect to the task at hand.

For this experiment, we needed a task that was familiar to all of our participants and that they would know required a combination of innate ability and rigorous practice. For logistical reasons, we needed a task that people could understand very quickly, that could lead to failure in a very short time, and that could result in unambiguous and objective success. Because the participants in the study were going to be college students, video games seemed an appropriate choice.

We began our experiment by having participants play a very simple video game: an elf, controlled by the user, had to navigate through an imaginary world in search of a yellow crystal. Unbeknownst to the participants, in fact, there was no yellow crystal! Thus, failure was guaranteed. After experiencing the frustration of defeat, participants received one of two types of feedback. Half of the participants heard an evaluation typical of someone with a fixed mindset: "You lost the game. People are born with a certain amount of video-game talent. You are not a talented video-game player." The other half of the participants heard a comment typical of a growth-mindset evaluator: "You lost the game. Video-game skills can be improved through practice."

To determine how the two types of evaluations would affect subsequent mindsets, we then presented the participants with one-paragraph descriptions of twenty other video games, half of which were described as easy and half of which were described as difficult. After reading each game's description, we asked participants to rate, via questionnaire, how difficult they thought each game would be and how interested they were in playing it.

➤ Results and Implications

Although participants' performance on the initial game was identical—consistently poor by design—after hearing either the fixed- or growth-mindset feedback, they had very different feelings about which type of games they wanted to play. Participants given a fixed-mindset evaluation were not interested in stretching themselves, preferring the easy games. In contrast, growth-mindset-evaluation participants welcomed the opportunity to strengthen their skills, indicating that they would prefer to play the hardest of the hard games.

Thus, your mindset (as reflected in your criticism) can lead evaluated people to stick to their existing strengths to avoid failure or to seek out challenges as a way of improving. When people receive criticism that reminds them of the importance of effort, they gain the benefits of a growth mindset. When you criticize their "inherent" attributes, it encourages a fixed mindset, which in turn makes it less likely that they will improve. Criticism that encourages one mindset or the other is so powerful that it can affect people's future choices and attitudes toward challenges, regardless of their original mindset.

Mindsets can also affect actual performance. In one study, researchers asked people to complete a management task on a computer. The task involved running a furniture company: participants had to allocate employees and decide how to best guide and motivate them. Throughout the exercise, participants were supposed to revise their decisions based on periodic feedback they received about employee productivity. Researchers primed one group of participants for a fixed mindset (telling them the task measured their underlying capabilities) and the other group for a growth mindset (telling them the task would help them develop their management skills through practice).

While both groups initially fell short of the high production standards the researchers gave them, those in the growth mindset improved over time. They used feedback to learn from their mistakes far more than the fixed-mindset group. As a result, those in the growth-mindset group eventually got their companies' productivity up to par, while those in the fixed-mindset group lagged behind.

Experiment:

Can Praise Be Anything but Positive?

We have seen that people receive both sincere praise and flattery very positively. Praise also tends to have mildly positive effects on behavior and soothes the body. However, it seems that whenever social scientists hear that there is something, such as praise, that makes life uniformly better, they have to find a way to screw it up.

I grounded the study in a common kindness: telling people, "This will be easy for you." We usually think phrases such as this build confidence—"No need to worry about doing well"—and are considered compliments—"While for others this would be hard, you certainly will have no problems." However, these statements also imply that the praiser believes people will do well based on who they are, not on their efforts. If people subsequently fail, will the fixed-mindset encouragement leave them with no one and nothing to blame but their own intrinsic deficiencies? In the case of criticism, I found that a fixed-mindset evaluation such as, "You do not have the talent to do well at this activity," discouraged participants from seeking new challenges. Could fixed-mindset praise also take away the joys of success?

For the study, I, along with Ph.D. student Yeon Joo, used a car simulator that had a voice system (much simpler than the one used in Japan but more than sufficient for our purposes). Before we had them drive, we pretended to assess participants' driving skills through a test given on a computer. We told participants that one of the best ways to assess people's strengths and weaknesses as a driver was the "Trail-Making Test" (although this test assesses neuropsychological difficulties, it in fact has no relationship to driving performance). The Trail-Making Test involves a set of twenty-four circles scattered randomly throughout a page on a computer screen. Each circle contains a different letter (ranging from A to L) or number (ranging from 1 to 12). We told participants to use the computer mouse to click on each circle in the pattern 1-A-2-B-3-C-etc. as rapidly as they could.

Participants were then told that they would be driving on three separate courses: a town, a highway, and a desert. We pretended to measure participants' driving skills to make it appear as though the computer had chosen an easy or challenging driving course for them based on this evaluation. Before they began driving, the car told half of the participants the following:

> This course is designed to be very easy for you. Your skills are very well suited to this course. That is, this course is meant to highlight your strengths as a driver. Even if you don't try hard at all, you will drive this course very well.

During the drive, the car would remind participants that the course should be easy for them, making comments such as, "There is a curve up ahead. You will have no problem handling it." The car told the other half of the participants the following:

> This course is designed to be very challenging for you. However, if you consistently work to your utmost ability, you will handle this course very well.

These drivers were reminded that with sufficient effort, they would navigate the course successfully: "There is a curve up ahead. If you focus intently, you will be able to handle it."

Although participants believed that the computer had tailored the courses based on their performance on the Trail-Making Test, all participants drove on identical courses with respect to route, the other cars on the road, when pedestrians appeared, the trees and buildings in the background, and weather conditions such as thick fog and slippery roads. Even though the car voice in the "easy" condition told the drivers that they could easily overcome the challenges of the courses, all drivers drove the same, very difficult courses—none of the participants finished the courses without at least one collision. After driving each course, we gave participants a questionnaire to determine how they felt about their driving experience.

➤ Results and Implications

The results of the study demonstrate the pitfalls of the phrase "easy for you." The "easy-course" drivers felt less fulfilled by the driving experience and felt that the driving was less enjoyable. They even took out their negative feelings on the voice of the car, which they described as more frustrating, confusing, difficult to follow, inaccurate, and unreliable than did the challenging-course drivers.

Previously, I suggested that praise never hurts. This does not mean, however, that all *types* of praise are beneficial. Telling people that they are "destined to succeed" before they attempt a new activity can make any failures crushing. Thus, fixed-mindset praise, meant to make people feel better, can actually make people feel much worse about their work and more negative about the person who praised them if it turns out to be inaccurate.

Can People Have Too Much Faith in Themselves?

According to Dweck, an epidemic of fixed-mindset praise started in the early 1990s, when many parents and teachers became focused on increasing self-esteem by constantly telling children how smart and talented they were. This mindset ironically has a negative effect on self-confidence as children face challenges and failures. For example, when Dweck asked fixed-mindset children why their parents would talk with them if they had performed poorly on something at school, they would respond with comments like, "They think bad grades might mean I'm not smart." In comparison, growth-minded students would respond, "They wanted to make sure I learned as much as I could from my schoolwork," and "They wanted to teach me ways to study better in the future."

In the workplace, the culture of praising also exists, with some employees who harbor a fixed mindset unable to take criticism and needing constant validation, recognition, and reassurance. The increasing number of managers with fixed mindsets exacerbates the problem. They do not support training programs because they believe workers' innate talent constrains the potential impact of education or practice on performance. They also do not give as much credit for improvement when deserved or critical feedback when needed for employees to grow and advance.

In sum, give growth-minded feedback to motivate people to choose challenging tasks and to confront their mistakes. Praise for taking initiative, completing a difficult task, learning new skills, and acting on criticism all encourage a growth mindset. Managers especially play a key role in creating an environment that encourages a growth mindset by giving feedback and support that praises learning and perseverance rather than inborn talent.

Experiment:
Judging the Judges

For the most part, this chapter has focused on the feelings and behaviors of the person being evaluated. What about the people who witness others being evaluated? Does seeing a third party praised or criticized elicit similar reactions to those that result when you yourself are being evaluated? I, along with Ph.D. student Jonathan Steuer, decided that the best way to answer this multifaceted question was to put participants in a scenario that frequently involves the evaluation of numerous people: teaching. This study was somewhat unusual in that we had computers serving the role of three different confederates.

We told participants that they would prepare for a test with the assistance of a tutoring computer. The computer tutor presented each participant with twenty-five facts (e.g., "According to a Gallup poll, 85 percent of respondents indicated that 'cheapness' is one of the most serious faults a person can have" and "The less wire in a computer, the faster it runs"). After reading each fact, the computer asked participants how much they knew about the fact on a scale from 1 to 9. This was supposedly an adaptive teaching computer, such that the more the participants claimed that they knew about a fact, the fewer additional facts they would receive on the subject. (We verified via questionnaires that all of the participants thought that the system did adapt.) In reality, all participants received the same twenty-five facts to ensure that everyone had the same experience.

After working with the tutoring computer, the participants were given a fifteen-question, multiple-choice test by a second testing computer. While the tutoring computer did not directly give away any of the answers, the testing computer's questions did seem related to the tutoring. For instance, one of the questions—"What percentage of people tip less than 15 percent at a restaurant?"—related to the fact about cheapness.

The participants then went to a third evaluator computer to complete an assessment of the tutoring computer's work. The evaluator computer went over each question, remarking on how well the tutoring computer helped the participant to answer the question correctly. Because we expected that the participants would be curious about their own performance as well, the computer told each participant that they had answered the same seven of fifteen questions correctly, regardless of their actual responses. This ensured that participants' reactions would not be influenced by their own performance or by distractions associated with not knowing how well they did on the test.

To explore how people's perceptions of the evaluator computer differed when it made positive versus negative comments about the tutor, half of the participants had a praising computer while the other half had a criticizing computer. For those in the praise condition, the evaluator computer favorably described the tutoring computer's performance twelve times out of fifteen; the other three times, it described the performance as moderately negative. (We included a few negative evaluations to make the assessments believable.) For example, after telling participants that they had given a correct answer, the evaluator then provided one of five different positive responses about the tutoring, such as, "The tutoring computer chose extremely useful facts for answering this question. Therefore, the tutoring computer performed extremely well." Even when participants (ostensibly) had provided an incorrect answer, the computer gave one of three positive responses about the tutoring, such as:

> The tutoring computer was constrained by the limited number of facts it was permitted to provide. Given more time, the tutoring computer would have offered very helpful facts. Therefore, the tutoring computer performed as well as possible for this question.

For those in the criticism condition, the negative evaluations paralleled the positive evaluations. For example, after telling participants that they had given a correct answer, the evaluator computer then provided one of five different negative responses about the tutoring computer, such as:

> The tutoring computer failed to provide useful facts for answering this question. Therefore, the tutoring computer performed poorly.

After working with the evaluator computer, participants filled out a paper-and-pencil questionnaire that measured their attitudes toward the tutoring, testing, and evaluation computers.

➤ Results and Implications

How did people feel about the different computers? Just as people dislike an evaluator who criticizes them, participants disliked the evaluator computer that criticized the tutor and liked the one that praised the tutor. Even though participants knew that the tutor computer had no feelings that could be hurt, the evaluator criticizing it led to negative feelings toward the evaluator.

People also thought that the computer that criticized was more intelligent than the computer that praised, even though the two versions of the evaluator's comments were equally complex in terms of grammatical structure and semantics. This is consistent with a classic study by the great social psychologist Solomon Asch, where he found that a person described as "intelligent and polite" was viewed as "wise" only 30 percent of the time, but a person described as "intelligent and blunt" was viewed as "wise" 50 percent of the time. In sum, you view someone who criticizes others more negatively than someone who praises, but you also view that person as more intelligent.

The evaluator computer's comments about the tutoring system affected not only how participants felt about the evaluator but also their perceptions of the tutor. When the evaluator computer praised the tutoring system, participants felt much more positive about the tutor than did people who witnessed the tutor criticized. For example, praise participants thought that the tutor was significantly more helpful than did criticism participants, although the information provided by the tutor and the participants' scores on the test were identical. Compared to criticism participants, praise participants also believed that the tutor computer contributed more to boosting their test score, both in general and relative to their prior knowledge. Even though the computer in reality ignored the participants' indications of how much they knew about each fact, participants who heard the tutoring computer praised considered it significantly more responsive to their prior knowledge.

Were these reactions because people believe that computers are always right? No: on average, the participants ranked the evaluator computer's judgments to be more *inaccurate* than accurate. Nonetheless, the mere existence of a positive or negative evaluation affected participants' perception of the tutoring computer's performance. The foregoing suggests that if you evaluate a person, it will change others' perceptions of that person, even when they are qualified to judge her or him for themselves. For example, if you praise your friend's performance, people will think that your friend did well, even though they know that your evaluation might be biased. Conversely, you can undermine virtually anyone's success by highlighting even a single deficiency. And saying, "It's just my opinion" doesn't obviate this effect. If you truly want people to make judgments for themselves (and to not judge you), keep your opinions to yourself.

* * *

The previous results have particular import for those who praise and criticize others for a living. While teachers and managers certainly fall into this category, it also includes professional critics of books, films, restaurants, cars, and other consumable products and services, whose raison d'être is evaluation. How do critics gain a positive reputation? Harvard Business School professor Teresa Amabile conducted an experiment to find out. She wrote two reviews of a nonexistent book. The reviews were identical except that at ten places in the document, she either inserted positive words (e.g., "successful" and "both interesting and engaging") or negative words (e.g., "unsuccessful" and "neither interesting nor engaging"). She then gave participants a questionnaire about the reviewer.

The results showed that participants saw negative reviewers as more intelligent and competent and as having more literary expertise than positive reviewers. This is consistent with the results from the previous experiment, in which participants saw the evaluator that criticized the tutoring computer as more astute than the one that praised. As Amabile puts it, "Only pessimism sounds profound. Optimism sounds superficial."

The presumptive intelligence of negative evaluators also occurs in the case of movie critics: critics who dislike most movies are seen as much smarter than critics who like most movies.* Surprisingly, people intuitively know that there is a perceived link between criticism and intelligence. Amabile showed that when asked to present in front of an audience that is described as having a higher intellectual status than the presenter, presenters became more negative.

In sum, critics, and all other evaluators, must decide whether they want to seem "clever and contemptible" or "kind and clueless." Thus, criticize only when it is urgent to do so or when you're trying to look smart.

---

* It is ironic that movie critics who hate movies are seen as smart; how smart is it to choose a profession in which you spend your time watching things you dislike?

Self-Evaluation

The ancient Greeks inscribed "Know Thyself" in the forecourt of the Temple of Apollo where the Oracle at Delphi resided. They knew that the ability to accurately assess yourself and to integrate evaluations from others (such as the Oracle) would reap great benefits. Today, this aphorism has essentially morphed into the much more risky, "Know thyself and tell others about it," as reflected in the dramatic growth of blogs and social networking sites. The notion of self-evaluation as a public rather than a private process has been institutionalized in the workplace since the 1960s via the requirements to both formally evaluate yourself as well as to formally respond to the evaluations of your bosses, peers, and subordinates.

How people perceive others' self-evaluations is complicated because self-evaluators' motivations are likely to be many-layered, conflicting, and nuanced, as they balance the urge to make themselves sound good against the expectation to be accurate. My approach of using computers as confederates made it possible for me to very cleanly investigate the issue, as I could determine people's feeling about self-evaluators without the difficulties that come with person-to-person interaction.

Experiment:

Let's Not Play the Blame Game

When you work with someone and failure occurs, should you be modest and blame yourself, or is it better to blame your partner? To answer this question, I needed a context in which a person would interact with a computer during an activity that would lead to many mistakes and failures. The mistakes had to occur frequently (to provide multiple opportunities for self- or other-evaluation) and in an obvious way.

Who caused the problem also had to be ambiguous enough that the computer could plausibly blame itself or the participant for it.

After pondering these requirements, I came up with the idea of using a voice-recognition system as the context. For a variety of reasons, these systems often fail at comprehending what people say, either misunderstanding what was said or simply failing to make any sense of it. For example, all too often an airline system thinks that you said "San Francisco" when you wanted "San Antonio"; a computer company thinks you want to make a "purchase" when you really want "service"; or a ticket reservation system thinks that the caller wants "noon" instead of "June." Perhaps even more commonly, a system finds it impossible to even guess what you meant, forcing you to repeat what you said (often multiple times).

When a failure occurs, the system must acknowledge the problem and then explain the reason for it—in other words, place the blame. Because breakdowns occur frequently and the computer drives the interaction through the questions it asks, this context presents a perfect opportunity to compare people's responses to how blame is placed.

I, along with Stanford undergraduates Armen Berjikly and Corinne Yates, built a telephone-based system for acquiring books via Amazon. The system allowed participants to search for books, listen to descriptions, browse the best-seller list, place books on a wish list or in a shopping cart, and make purchases. To facilitate the experiment, we had all of the participants inquire about the same products in the same order, and we ensured that each participant experienced misrecognitions at precisely the same points.

We created two versions of the software, identical except that they employed different methods for handling blame. The first version took the approach of almost all real-world systems by criticizing itself, saying for instance, "This system did not understand the selection. Please repeat it." This is self-evaluation, as the system modestly blames itself for the misunderstandings. The second version blamed the other obvious candidate: the user. In this situation, the system offered a response such as, "You are speaking too quickly. Please repeat your response," or "You must speak more clearly. Please say it again."

After using the system, participants responded to a questionnaire that asked how much they liked the interface, how willing they were to buy the various books the system had presented, and how competent they thought the speech-recognition system was.

➤ Results and Implications

Participants strongly liked the modest system that criticized itself and hated the system that blamed them. The system that blamed the participant was also a terrible salesperson: participants were much less willing to buy books from it than the system that blamed itself. That is, participants were angry not only with the system that criticized the user but also with the company, refusing to buy its books.

The most interesting result came from the perceived competence of the two systems. Although participants were clearly very negatively disposed toward the system that criticized them, they actually thought that it was much more competent and made fewer recognition errors! This was despite the fact that the two systems made identical mistakes at identical points in the interaction. In sum, modesty undermines your perceived intelligence so much that even insulting the person you are working with makes you seem more competent to that person than criticizing yourself.

Experiment:

Enough About Me. Let's Talk About You. What Do You Think About Me?

In the previous experiment, the participant had a stake in the criticism. When things went wrong, participants obviously did not want to be blamed—they may have welcomed the computer's modesty simply because it let them off the hook. To ensure that the conclusions about modesty were robust, I decided to examine a situation in which the participant could be more objective. This would allow us to compare and contrast the advantages and disadvantages of being seen as a braggart (self-praiser), a lauder (other-praiser), a self-deprecator (self-criticizer), or a critic (other-criticizer)

The most straightforward way to make all of these comparisons simultaneously was to expand the earlier experiment involving tutoring, testing, and evaluation. In the original experiment, participants heard an evaluating computer either praise or criticize the tutoring system. For the extension of the experiment to include self-evaluation, I had a new set of participants go not to the third, evaluating computer but instead go back to the same computer that tutored them to hear it either praise or criticize itself. These self-evaluation conditions were identical to the other-evaluation conditions except that the tutor computer referred to itself as "this computer" rather than "the tutoring computer." (The computer did not refer to itself as "I" because we feared that this would seem odd and overly anthropomorphic.)

➤ Results and Implications

How do people who are not involved in an interaction feel about someone who is modest, that is, a self-criticizer? Although generally people like those who criticize less than those who praise, participants liked the computer that criticized itself much more than the computer that praised itself. Furthermore, participants also liked the computer that criticized itself much more than the computer that criticized another computer. Jack Benny was right when he bragged, "Modesty is my best quality."

However, as far as perceived competence, modesty was again a poor strategy: consistent with the Amazon study, participants felt that the computer that criticized its own performance was less competent than both the computer that praised itself and the computer that criticized the other computer. Thus, the use of modesty involves another "kind but clueless" trade-off: while a laudable quality, modesty, unfortunately, is also very convincing.

How do people feel about those who praise themselves versus those who praise others? In this case, no trade-off exists. Participants liked the computer that praised itself much *less* than the computer that praised a different computer. Participants also felt that the tutoring computer that praised itself was less competent than the tutoring computer that was praised by another, evaluator computer. This makes the choice between self-praise and other-praise clear: never praise yourself when you can have someone else do it for you. Thus, the best strategy in the workplace is a "mutual-admiration society": you praise your colleague, and he or she praises you back. This will lead to both you and your colleague seeming smarter and more likeable than if you each praised yourself.

This strategy can readily be used when someone introduces you before a presentation. Whenever I am asked to speak, I make sure to know my introducer's name and position. After the typically positive introduction, I say, "Thank you for the kind introduction. It is a particular honor to be introduced by [name of person], as she is an outstanding [name of position]." By doing this, my audience admires both me and my introducer more: the praise from the introducer inflates my perceived competence, and my praise of the introducer inflates her or his perceived competence. This also leads to a positive spiral, as the increased perceived intelligence of the introducer makes people feel that her or his positive comments about me are even more valid.

If you want people to like you and don't care how smart you seem, criticize yourself and praise others. If you want to seem smart and don't care about being liked, than criticize others and don't be modest. However, adopt the latter conclusion with caution because if people do not like you, they will think you are competent *but will not describe you positively to others or reward you for your competence.* While your criticism will influence them, you will gain a reputation not for excellence but for unpleasantness. And, of course, don't directly criticize the person you are interacting with when you can criticize a third party.

- Praise others (but not yourself) freely, frequently, and at any time, regardless of accuracy. Emphasize effort over innate abilities. When possible, establish a mutual-praise agreement in which you and a partner praise each other.

- Criticize others with caution, keeping it brief and specific, and always with clear follow-up actions. Present ways to improve and resolve the criticism, and emphasize the importance of effort for success. Afterward, give people time to process and to respond when they are ready.

- When mixing praise and criticism, offer broad praise, brief criticism focused on specific steps toward improvement, and then lengthy and detailed positive remarks.

- Modesty might win you friends but will also be believed, so only criticize yourself when it is accurate and constructive to do so.

- If you want to seem competent, then reverse the previous advice: praise yourself, criticize others, and don't criticize yourself.

CHAPTER 2

Personality

As a consultant and a speaker, I constantly meet new people. I often find it daunting to keep track of who they are as well as how to treat them. I used to manage this by describing them to my son, Matthew. I would think of a single trait that stood out about each person I wanted to remember and amplify it.* Thus were born a number of characters: "Numberer," who started every comment with "first" and then moved on to "second" and "third"; "Kennel," who used dog references whenever possible ("every dog has his day," "he's a pit bull," "she's like a dog with a bone");† "Rainbow," who would always write on the board with at least ten different colors; "Balloon," who pronounced words that started with "wh" with an enormous exhale; and "Animator," who filled his PowerPoint slides with dancing stick figures and spiraling text.

While this method amused Matthew, it proved less effective for

* My "embellishments" have caused Matthew some troubles. When I took him to his first children's musical, *The Adventures of Tom Sawyer*, he asked me, "Why is everyone singing?" I told him that the story was set in the mid-1800s and that in those days people sang instead of talked. I considered this a healthy way to encourage his imagination until I found out that while studying the Civil War at school, he had demonstrated his "knowledge" of the antebellum period to the class by describing the unique form of communication popular at the time.

† I tried to get him to change, but you can't teach an old dog new tricks.

helping me remember people professionally. For example, when I later bumped into "Kennel," it was all I could do not to laugh at each new dog metaphor. While every tidbit of information about a person can be helpful at some point, people do not have enough brain power to leverage each of those facts at the right moment. Concentrating on certain "unique" traits meant that I ignored or forgot other characteristics that might actually help me get along better with someone.

Rather than focusing on the ludicrous, social scientists have found that many characteristics appear in systematic groupings across people and reliably predict their attitudes and behaviors; they call these clusters "personality traits." For example, people who are aggressive also tend to be assertive, forceful, outgoing, and competitive. If a person has one of these characteristics, you can almost certainly expect the rest. Furthermore, the effects of personality traits tend not to change across situations. Thus, if you are outgoing at work, you probably tend to be outgoing at parties, in restaurants, and at the grocery store. Personality traits are remarkably stable across time as well: according to a review of more than 150 studies by psychologists Brent Roberts and Wendy del Vecchio, people's personalities are essentially set by the time that they are five years old. This explains how you can encounter someone after twenty years and exclaim, "I knew you'd say that!"

In general, if you identify a few personality characteristics of a person, you know a great deal about how she or he will think about and respond to other people and how you should think about and respond to her or him. Thus, you can use people's personality information (as opposed to someone's obsession with dogs or affinity for colored markers) to guide how you should interact with them. For example, imagine that you are pitching an idea to a colleague who is shy, reserved, and tentative. You, on the other hand, are assertive, dominant, and forceful. How should you proceed?

Four different adages could guide your behavior. The first calls for you to "Be true to yourself." You have a certain way of doing things: by this logic, you should act naturally, regardless of the other person's particular characteristics. If you try to adjust your behavior, you will not be at your best and might seem insincere.

A second strategy dictates that you should act according to your role in the current situation: "Cobbler, stick to your last [trade]." People generally think of effective salespeople as friendly, so because you are making a pitch, you should act friendly. Neither your colleague's reticence nor any other of your or your colleague's characteristics should influence how you present yourself. In other words, you should act according to the situation and your role in it, regardless of your personality or those of the people around you.

The third strategy, "Birds of a feather flock together," is known in the social science literature as "similarity-attraction." This is the idea that the more similar two people are, the more probable it is that they will like, trust, and respect each other. Conversations and collaborations will go more smoothly because similar people can predict one another's reactions by simply thinking, "What would I do?" In this case, you would present your idea in a calm and restrained manner to better match your colleague's quiet personality.

The final possibility is to assume that "opposites attract," known in the social science literature as "the principle of complementarity." This position argues that when people encounter someone different from themselves, an energy and fascination draw them together. Social relationships work best when one person's weaknesses complement another person's strengths. Thus, with your quiet colleague, you should act boisterous and exuberant: your opposing personality will intrigue him and attract his interest.

To investigate which of these viewpoints is correct, my lab and I conducted a series of experiments. Studying personality using people alone proves problematic because every person comes with many characteristics—such as gender, age, and appearance—that are independent of the traits under investigation. For example, if I wanted to compare how people react to cold people as compared to friendly people, the lab would need two confederates who are essentially identical in every respect other than their friendliness (or a superb performer who could convincingly and unambiguously seem alternatively friendly and cold). They would have to behave alike, look alike, and sound alike in every other way. Or I would have to tell participants, "Ignore everything else about this person except how friendly she is" (which would likely encourage them to think about those other characteristics).

Once again, computers had opened up an entire world of possibilities.

Billions of People, Four Personalities

People use literally hundreds of terms to casually refer to personality: type A, easygoing, charismatic, adventurous, self-centered, "wild and crazy," and so on. However, social scientists have found that out of all of these, only two critical distinctions determine how people interact with each other (the vast majority of the other traits describe how people approach life in isolation). As a result, you need only two questions to characterize and guide your interactions with any person.

The first question is whether someone is extroverted or introverted. Extroverts become engaged and excited by other people, especially getting energy from large groups. The "life of the party," extroverts don't wait for conversations to start: they initiate them and keep them going. At work, extroverts prefer group projects to individual assignments and like interacting with clients rather than with data. On teams, an extrovert pays attention to other individuals and takes their feelings into account. They enjoy jobs that encourage bold and quick decision making. As managers, they are charismatic, energizing their team with their words and actions.

In contrast, introverts prefer "alone time" to socializing with others. Private activities such as daydreaming and reading invigorate them, while it drains them to interact with people. Introverts listen more than they talk in a conversation, taking time to think before they respond and developing their ideas by reflecting privately. In the office, introverts prefer working alone and being responsible for things and information rather than people. Introverts can more readily focus their attention and keep it focused, contemplating all angles of a problem. They dislike jobs that require decision making, especially decisions they must make without full information. As managers, they tend to direct people "by the numbers" and treat all of their subordinates equally.

Extroversion and introversion are opposites; the other two personality types are also opposites: critics and sidekicks. "Critics" don't enjoy spending time with people and look for opportunities for schadenfreude, pleasure derived from the misfortunes of others. When they do interact with people, they quickly make their negative feelings and judgments clear. Critics do not listen well and focus more on themselves and their own feelings than on those of the people they talk with. They tend to use few but very strong words.

At work, critics prefer one-on-one interactions and to have the upper hand; they make particularly poor subordinates. They truly dislike jobs that require kindness to the public; they do not suffer fools gladly. Like extroverts, they enjoy jobs that encourage bold and quick decision making, but unlike extroverts, they prefer that those decisions have nothing to do with people. As managers, they motivate others through shouting and criticism.

The opposite personality of the critic is the "sidekick." They tend to think very optimistically and see the best in others. Sidekicks enjoy interacting one-on-one, but large groups can overwhelm them. They like to talk about people and listen better than any other personality type. Sidekicks tend to be verbose and remarkably expressive, but they ask as many questions as they make statements.

At work, sidekicks enjoy working underneath one person and not necessarily on teams. They like executing on the decisions of others but dislike decision making, especially when those decisions impact many people. Sidekicks rarely work as managers, as they prefer neither to direct nor evaluate people. They do well with service jobs as long as they are not high stress.

Another way you can think about the four personality types (extrovert, introvert, critic, and sidekick) is in terms of two dimensions: control and affiliation. The "control" dimension distinguishes dominance versus submissiveness. Dominants—extroverts and critics—have a strong desire to control or influence others; submissives—introverts and sidekicks—try to avoid making decisions for others. The "affiliation" dimension divides people who are friendly and like to interact with and share feelings with others—extroverts and sidekicks—from people who are cold and try to avoid interactions and revealing feelings—introverts and critics. That is, extroverts are dominant and friendly, critics are dominant and cold, introverts are submissive and cold, and sidekicks are submissive and friendly, as shown below.

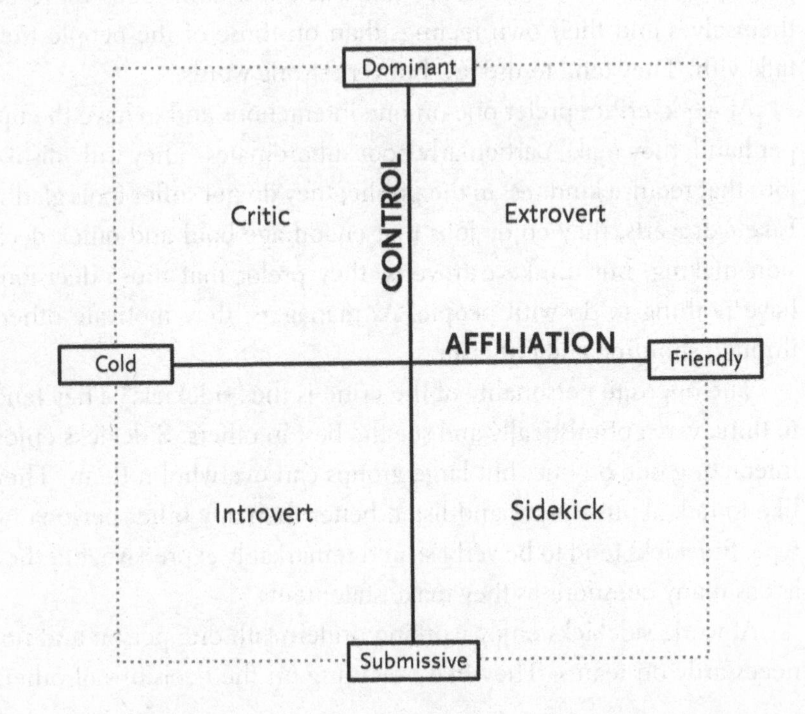

Given that everyone has one of these four personalities, it becomes straightforward to figure out the most effective strategy for working with a particular personality type. In this study, working with Ph.D. student Kwan Min Lee (now a professor at the University of Southern California), I focused on how introverts and extroverts work with other introverts or extroverts. Specifically, I paired extroverted or introverted participants with extroverted or introverted computers to determine which of the four adages—"Be true to yourself," "Cobbler, stick to your last [trade]," "Birds of a feather flock together," and "Opposites attract"—would prove most accurate.

Experiment:

Too Much Talk, Too Little Talk, or Just Right?

First, I looked for a context that wasn't based too obviously on personality. I also needed participants to feel that it was natural to encounter a variety of personalities in that context (both extroverts and introverts).

After some deliberation, I decided that the online auction site eBay provided the perfect venue for the study. While some personality types love the excitement of auctions and others buy only the "fixed-price" items, virtually everyone has something to sell or something to buy: thus, all personalities are represented. And while descriptions of items seem far removed from issues of personality, they actually reveal a great deal about the seller. People with dominant personalities write descriptions with confident assertions and rich detail, and friendly people refer to themselves and others with phrases such as "I am sure you will like this." Thus, an extrovert (a dominant and friendly personality) might describe a lamp on eBay as follows:

> This is a reproduction of one of the most famous of the Tiffany stained-glass pieces—the colors are absolutely sensational! This first-

> class, handmade copper-foiled stained-glass shade is over six and one-half inches in diameter and over five inches tall. I am sure that this gorgeous lamp will accent any environment and bring a classic touch of the past to a stylish present. It is guaranteed to be in excellent condition! I very highly recommend it.

Introverts' descriptions look completely different. Showing their submissive side, their descriptions tend to say as little as possible. They also tend to avoid descriptive language, instead primarily using nouns and verbs. Even when presenting the facts, introverts take a tentative tone, as they lack confidence. Manifesting their coldness, introverts prefer to talk about things rather than people and keep themselves, their feelings, and their audience out of the conversation. An introvert might describe the same lamp as follows:

> This is a reproduction of a Tiffany stained-glass piece. The colors are quite rich. The handmade copper-foiled stained-glass shade is about six and one-half inches in diameter and five inches tall.

For this study, we needed participants who were clearly extroverted and others who were clearly introverted. A few weeks before the experiment, we gave more than a hundred people a personality assessment. We asked people how well statements such as the following described them: "I spend my leisure time actively socializing with a group of people, attending parties, shopping, etc." (extrovert statement); "I enjoy having a wide circle of acquaintances" (extrovert statement); "After prolonged socializing, I feel I need to get away and be alone" (introvert statement); and "I am a person somewhat reserved and distant in communication" (introvert statement). Based on the results of this assessment, we invited the most extroverted and most introverted people to participate in the study, taking care not to tell them how we chose them.

Once participants arrived at the lab, we directed them to an online auction site that had an eBay-like interface. On the site, nine antiques were up for auction, including a stained-glass lamp (described above), a limited-edition Marilyn Monroe watch, a Russian circus poster, and a U.S. Treasury award medal. We selected items of limited desirability to the average person so that participants' intrinsic interest in particular items would not affect the results of the study.

We wrote an extroverted and introverted version of each of the descriptions, just as in the lamp example. Half of the extroverted participants and half of the introverted participants saw only the extroverted descriptions; the other half saw only the introverted descriptions. We made sure that regardless of the style of description, the two versions—which included differences in length, word choice, phrasing, and references to self and other—presented identical factual content. This ensured that quality would not factor into the participants' decision making.

After participants read the description of each item, the site asked them to indicate how much they would pay for the item, how much they liked the description, how they felt about the auction, and how they felt about the seller who wrote the description. Thus, I could determine which, if any, of the adages outlined above were correct. If personality clarity is most important, all participants would react similarly to both types of descriptions because they both unambiguously reflected a specific personality type; that is, each description was "true to itself." If a "pitchman" is always supposed to be "exciting and compelling," then all participants would like the extroverted descriptions because that personality fits a traditional salesperson. However, if the match or mismatch between the buyer's and seller's personality types is more important than whether the seller is of one type or another, the extroverted participants would like the extroverted descriptions more (similarity-attraction) or like the introverted descriptions more (complementarity), and vice versa.

➤ Results and Implications

The results clearly indicated that similarity between the buyer and seller leads to the most successful interactions. Extroverts who looked at an item described in an extroverted way indicated they would pay more for it, even though the introverted version described the identical item. Introverts exhibited the opposite preference: their questionnaire results showed that they would pay more for the items described in an introverted way. Similarly, extroverts preferred the extroverted descriptions and introverts, the introverted. Thus, despite both sets of descriptions containing the same facts about each item, the personality reflected in the descriptions strongly influenced potential buyers based on the buyers' own personalities.

A review of the social science literature further confirms that people prefer those with similar personalities to themselves. Sharing a personality trait predicts success in roommate pairings, dating, and long-term relationships such as marriage. People even get along better with strangers who are similar to them. In general, when people encounter a person with a personality similar to their own, she or he will be considered likeable, intelligent, and trustworthy. In fact, my Ph.D. student Youngme Moon (now a professor at Harvard Business School) and I concluded, based on a review of the research literature, that people will attribute almost any positive characteristic to people who share personality characteristics with them. Similarity-attraction affects people to such a degree that they feel positive toward not only similar people but also anything associated with those similar people. For example, in the experiment, not only did participants like the sellers who were similar to themselves, they also felt more positive about the items associated with the similar sellers.

While I accepted the findings for similarity-attraction, they did initially feel counterintuitive to me. When I read the eBay descriptions, the extroverted ones clearly seemed more exciting and friendly. Wouldn't that make them more persuasive to everyone? It wasn't until designing another personality experiment that I really understood the basis for similarity-attraction.

My Ph.D. student and I (both dominant personalities) were creating content for a computer interaction partner to present in either a dominant or a submissive tone. For example, at one point, the dominant computer would say: "You should definitely select option A instead of option B. There are at least six reasons why this is the right option. I am 90 percent confident of this assessment." The parallel submissive text read: "Perhaps you should select option A instead of option B? It seems like there are reasons why this might be the right choice. I am 40 percent confident of this assessment."

After reading the two versions, we looked at each other and knew instantly we had had the same thought: this study couldn't possibly work! How could those pathetic, tentative comments persuade anyone? If the computer didn't even trust itself, how could anyone else be expected to trust it? We were deep in discussion over what to do when another of my students (who has a submissive personality) walked by the door of the lab. We ran out, pulled him inside (fortunately, submissive personality types don't complain much about this type of treatment), plopped him down in front of the two sets of recommendations, and asked, "Which do you like better?"

After some hesitation, he said, "These," pointing to the *submissive* descriptions. "These comments sound thoughtful. This person seems to know when he is unsure and isn't ready to conclude something, so I feel like I can maybe trust his judgment." We were both flabbergasted. "Those," the student continued, looking at the dominant comments, "are like getting advice from a bull in a china shop. This person sounds kind of full of himself. Maybe he's talking without thinking? I don't think I can take those comments as seriously."

The submissive student's radically different interpretation of the comments made me understand similarity-attraction. Personality fundamentally affects how you see the world, and so people with different personalities see it very differently. This makes it difficult for opposite personalities to understand each other's point of view. On the other hand, when you encounter people who think and feel just like you, you naturally understand and feel affinity with them and what they say.

In addition to the thoughts and actions of a similar personality being more understandable, they positively reinforce who you are. When someone thinks or behaves the same way that you do, it confirms that your approach to life is the right one. Conversely, an incompatible personality can feel like a challenge or a threat; it subtly implies that your approach to life is wrong. Accordingly, my dominant graduate student and I instantly thought the dominant comments were universally persuasive and were puzzled that *anyone* could find the namby-pamby comments convincing. When we said something, we meant it. From our perspective, what was the point of talking if you did not announce the right answer? My submissive student, conversely, spoke cautiously and couldn't imagine why someone would prefer the "overly confident" comments.

Similarity-attraction arises not only because it is ego supporting; evolution also has a role in its origins. Throughout most of history, people lived in communities with individuals who shared languages, customs, and a particular orientation to the world and to each other; different communities had different languages, customs, and orientations. As a result, the more a person behaves similarly to you, the more likely that he or she came from the same clan and the stronger the incentives for mutual survival. Thus, people evolved a very simple notion of personal ties: "If these people behave similarly to me, they must be personally tied to me, and if they are personally tied to me, they deserve to be viewed and treated as friends."

While you can usually determine someone's personality from listening to what he or she says, people also display their personality type in other, more subtle ways. As discussed in a book on designing pictorial computer characters by New York University professor Katherine Isbister (once my Ph.D. student), personalities can come across through posture and gestures. Extroverts fill the room with big gestures and have a wide stance, an erect body, broad facial expressions, and make a great deal of eye contact. In contrast, introverts hold their bodies tight with very limited gestures, keeping their heads down, shoulders slumped, and eyes averted to avoid eye contact. Extroverts tend to lean toward people with whom they are interacting, while introverts tend to lean away.

While people don't usually think about the connection between personality and people's voices, the way one speaks is surprisingly consistent and revealing. Vocal indicators are so powerful that you can quite accurately judge people's personalities based on them (as demonstrated in my book *Wired for Speech*) even when the person is speaking in a foreign language (with the exception of tonal languages such as Chinese). In the next study, Kwan Min Lee and I investigated whether people are so attuned to personality that when it is merely reflected in people's voices, and not in what they are saying, it will nonetheless generate similarity-attraction.

Experiment:

If You Say "Tomato" Like I Say "Tomato," We've Got Ourselves a Deal

To make sure that the *only* difference participants heard in the voice in our experiment was extroversion versus introversion, we used a simple, computer-generated voice. To make the voice sound extroverted, we increased its volume, frequency range, and speech rate (traits researchers have established as markers of extroversion). To make the voice sound introverted, we used the same voice but lowered its volume, frequency range, and speech rate. Before the actual experiment, we tested the voices with people who were neither strongly extroverted nor introverted to check that people did not generally find one voice more attractive, credible, informative, or persuasive than the other. As before, we also used a personality test prior to the experiment to select equal numbers of clearly extroverted and clearly introverted participants for the study.

For the context of the experiment, we chose a bookselling Web site that looked like Amazon. Synthetic speech seemed very natural for such a site because having so many books and a rapidly changing inventory would make it impossible to use a recorded human voice to present every book review. The Web site displayed the cover, title, and author of five different books in a manner similar to Amazon. To get a review of the book, participants clicked a link to an audio file that read the description using the synthetic voice previously described. Half of the extroverts and half of the introverts heard all the reviews read by the "extroverted" synthetic voice, while the other half heard the descriptions read by the "introverted" synthetic voice. We kept the content of the reviews (the scripts) identical for each book and varied only the qualities of the voice.

After playing the review of a given book, the site asked participants how likely they would be to buy the book. After participants went through all of the book reviews, we gave them a questionnaire that asked their feelings about the experience, the voice, the reviews, and the reviewers.

➤ Results and Implications

Birds of a feather once again flocked together: extroverts liked the extroverted voice more, while introverts liked the introverted voice more. Even though the person who actually reviewed the book clearly was not the one speaking in that robotic voice, participants liked the reviewer more when the voice that read the review matched their own personality. Matching participants also rated those reviewers as more credible. Even though the reviews were identical, extroverts liked them more when read by the extroverted voice, while introverts preferred them when read by the introverted voice.

Would people put their money where the computer's mouth was? I found that similarity-attraction did indeed guide buying behavior. Extroverted participants were more likely to buy the books when an extroverted voice presented the reviews, while introverted participants were more likely to buy when hearing the reviews read by an introverted voice. Not surprisingly, introverts (who enjoy solitary activities such as reading) were more likely to buy the books in general: personality type predicts people's behavior in isolation as well as when interacting with others.

My lab has since completed many other experiments in a variety of contexts that demonstrate the ubiquity of similarity-attraction with respect to personality. While working on a cooperative task, people gave a computer more credit for success and less blame for failure if the personality manifested in the language of the computer matched the personality of the person (counteracting their self-serving bias). Even a hint of a personality is sufficient to influence overall perceptions: Harvard professor Youngme Moon showed that when the *introduction* to a computer-based "Entertainment Guide" matched users' personalities, users found the recommended music to be significantly better, even though the recommendations themselves were identical.

My research clearly shows that no single personality type is "best." Despite the fact that archetypal personalities for different roles exist (domineering bosses, introverted programmers, friendly salespeople), the most effective personality to have in any given situation depends on whom you are interacting with. Getting along with someone is not just a matter of being "likeable" or "appropriate"; whether you share personality traits makes the difference.

Personality influences how we view people all the time, not just during popularity contests. Although the eBay and Amazon experiments did not explicitly draw participants' attention to the computer's personality, it nonetheless influenced them in remarkable ways. Personality affected not only how people judged the computer but also how they judged the items the computer described. Thus, even when people have full access to "objective" information, personality plays a role in their perceptions and decisions.

Sometimes these effects can be insidious. Similarity-attraction can result in hiring that amplifies the existing uniformity of a work team. For example, in a job interview, when an interviewee with a similar personality to your own gives a "good" answer, it may just seem good because it is what you would say, said in the way that you would say it: similarity is ego supporting.

"Thin slices," a phenomenon discovered by psychologists Nalini Ambady and Robert Rosenthal, are strongly held impressions of a person after a very short interaction. They have been lauded as an extremely effective and rapid way to assess others, but these snap judgments can be overly connected to a sense of similarity. While too much introspection can make your judgments less accurate, don't be a "sucker" for similarity and ignore other deficiencies. A healthy suspicion of similarity can help diversify the workplace and remove blind spots.

The power of similarity and the hazards of alluring similarities raise some unsettling questions. Should I try to befriend and work with only people similar to myself? Are all other relationships doomed to mistrust and failure? As a friendly and dominant person, for example, what should I do when I work with introverts, critics, or sidekicks?

My first idea was to tone down all the telltale signs of personality. Why should I talk in a strongly extroverted way and drive away introverts if I could seem neither extroverted nor introverted and thereby alienate no one? Unfortunately, in a classic paper, psychologist Solomon Asch demonstrated that we prefer people with a clear personality—ambiguous personality manifestations are disliked by everyone. You become like Charlie Brown, whose personality was so neutral that the only thing that could be said of him was, "Of all the Charlie Browns in the world, you are the Charlie Browniest."

I then thought of trying to appeal to everyone. For example, I could manifest one personality with the way I talked to appeal to some of the people in a group, and use nonverbal cues, such as posture, to match the personality of other group members. That way, everyone could find something to like. I investigated this idea by using computer agents. Specifically, Katherine Isbister and I had participants interact with stick figures whose postures suggested extroversion or introversion and who had word balloons that incorporated extroverted or introverted text. When we mixed and matched the postures and the text, we found that people preferred the pictorial agent whose language indicated the same personality as their body posture did, regardless of what their own personality was, and thought the consistent stick figures were more intelligent, fun, and persuasive.

Other research further confirms the unpleasant dissonance of multiple personalities. Social psychologists Nancy Cantor and Walter Mischel showed that if someone is described with a set of conflicting adjectives, people like the person less. Psychologists James Graves and John Robinson II showed that when a guidance counselor's language mismatched his body language in terms of extroversion versus introversion, clients rated the inconsistent counselor as less genuine and trustworthy; they also sat farther away from him.

Why is an inconsistent personality so off-putting? For starters, it requires your brain to work harder as it tries to resolve the inconsistencies. You end up drawing on limited cognitive resources to figure out what is going on: "Does he want to continue the conversation or not?" and "Am I supposed to make the next move, or is he?" This confusion makes it harder for you to focus on the task at hand.

People also dislike inconsistent personalities because they associate inconsistent personality manifestations with untrustworthiness. People correctly assume that bodies and words are naturally linked. To experience this for yourself, hold your arms wide, put a huge smile on your face, and loudly say, "Perhaps you might want to do this?" For the vast majority of people, this is extremely difficult—the extroverted gesture does not mesh with the introverted tone of the suggestion. Thus, when we observe these deviations, we assume that the speaker is manipulating how she or he communicates.

The repulsion of inconsistency was most clearly brought home to me when I heard Mike Tyson speak on television for the first time. Tyson, the onetime heavyweight boxing champion of the world who bit off the ear of one of his opponents and spent years in prison for rape, loomed on the screen. He was the epitome of a dominant/cold person. However, when he began to speak, I was totally startled: he had a soft, high-pitched voice with a slight lisp. If I hadn't known what he looked like or about his checkered history, I would have sworn that he was a gentle sidekick rather than a ferocious villain. Ironically, rather than softening his image, his unambiguous dominance and coldness coupled with a voice that suggested friendliness and submissiveness made him seem even more bizarre and repugnant to me than if he had simply had the deep, gruff voice that one would have expected.

So neither a neutral personality nor displaying multiple personalities simultaneously would have overcome the fact that my personality naturally distanced me from different personality types. I then wondered what would happen if I adjusted my personality to become more similar to the person with whom I was working. Would the initial impression of my personality overpower later attempts to act more similarly, leading me to appear "grossly ingratiating"? Or would becoming more similar garner all the benefits of similarity-attraction when it seems to occur naturally between people?

Experiment:

Is Imitation Flattery?

Youngme Moon and I wanted to create a situation in which a participant would work with an interaction partner with a clearly different personality that at some point would change to become more similar. For comparison, I used two other types of partner: one that would remain similar (the traditional similarity-attraction situation) and one that would remain different (staying "true to itself" despite the dissimilarity).

For this experiment, I focused on a different personality dimension from the eBay and Amazon studies: dominance versus submissiveness. In the first round of interaction, half of the participants worked with a computer with a similar personality and half worked with an opposite personality. We gave participants a modified version of the "Desert Survival Situation" (licensed by Human Synergistics International), a classic cooperative task researchers use to study attitudes and behaviors in interactive situations. In this task, participants were asked to imagine that they were in a plane that crashed in the desert. The plane had ten salvageable items, including a large knife, a compress kit, a flashlight with four batteries, and a box of kitchen matches. Participants were asked to rank the items in importance for survival, with 1 being the most important and 10 being the least important. We told them that to do well they should discuss their answers with their partner (the computer) before making their final choices.

The Desert Survival Situation has become extremely popular in social science research. Participants think it is fun and interesting; conversations are easy. The task also has the advantage that answers are not obviously wrong or right—even survivalists disagree. For our purposes, it was ideal because we could plausibly script the computer's random responses without the computer having to understand what the person had said. Finally, this task captures the most important characteristic of decision making in organizations: no one knows all relevant information, but everyone has something to contribute.*

When participants began the task, they entered their initial rankings of the items into the computer. The computer then provided its rankings, displaying them next to the participants': we arranged for the computer's rankings to be quite different from those of the participants. The participants and the computer then discussed the appropriate ranking of each item, with the computer displaying its comments on its screen and the participants typing their responses to the

* The task isn't perfect, though. When we tried to do an experiment at Kyoto University using a variant of the Desert Survival Situation, the participants struggled: we found out that Japan has no deserts! (Human Synergistics International has not had the same difficulties). Having learned that, we now use the "Moon Survival Situation" for cross-cultural experiments to level the playing field: few can claim personal experience with the lunar surface.

computer. After exchanging comments about the ten items, participants entered their final rankings. I then had them fill out a paper-and-pencil questionnaire that asked a variety of questions about how they felt about the computer, the interaction, and themselves.

To manifest different personalities while discussing the rankings, the computer varied its language style to suggest dominance or submissiveness. For example, the computer used strong language with frequent assertions and commands to express dominance. Conversely, the computer used weaker language in the form of questions and suggestions to express submissiveness. In addition, the dominant computer commented on each item before the participant; the submissive computer, after.

For each item, the computer gave one of two responses—one if it ranked the item higher than the participant had and one if it ranked the item lower. For example, if the dominant computer had rated the compress kit higher than the participant did, it would say:

> You should definitely rate the compress kit higher. The desert environment will certainly cause some kind of injury. The compress kit will also help close up serious wounds and reduce the risk of infection. You should move the compress kit up in the rankings!

If the submissive computer had rated the compress kit lower than the participant, it would say:

> Maybe the compress kit should be ranked lower? It seems somewhat unlikely that there are many ways to get cut in the desert. I think that the desert is actually one of the most sterile environments on earth. Could it be that the compress kit should be moved downward?

To determine the effects of changing personality versus keeping it consistent, we had participants complete a second round of the Desert Survival Situation with the same computer but a new set of items. For half of the dominant participants and half of the submissive participants, the computer radically changed its personality between the first and second round. That is, half of the participants who worked with a computer sounding dominant in the first round found that the computer became submissive in the second round, and for the other half the computer retained the same dominant speaking style, high confidence, and initiation of opinion in the second round. Similarly, for the participants with whom the computer started out submissive, the computer became dominant in the second round for half of them and for the other half it stayed submissive, retaining the same questioning and tentative speaking style, low confidence, and post hoc response to the participants' comments.

➤ Results and Implications

Before focusing on the issue of change, I checked to make sure that similarity-attraction holds in the case of dominance versus submissiveness by looking at the results for the participants who worked with a single-personality computer. Would a duo of leaders or a duo of followers really get along better than a leader and a follower? The results were striking and clear: similarity-attraction trumps complementarity. Although they received identical content, participants found the computer when its personality consistently matched their own more intelligent and insightful than when the computer's personality consistently mismatched their own. Similarity led participants to like the computer significantly more and to find the interaction more enjoyable and engaging, according to the questionnaire results.

I then turned to the question of whether, if you have a dissimilar personality to a working partner, you should stay true to your own personality or change to be more similar. Does change indicate charming flexibility or despicable ingratiation? The effects were enormous and the conclusion inescapable: people *love* it when you adopt a similar personality to their own. The computer when it changed its personality to match the participant's seemed much more intelligent, helpful, useful, and insightful than when its personality remained mismatched, and participants enjoyed working with it more.

When I saw how positively participants felt toward the computer when it became similar, I realized that I should compare those effects to similarity-attraction. Could being different and then becoming similar to a person be better than consistently being similar? When I analyzed the data, it confirmed my supposition: people felt that the computer that changed to be similar was more intelligent and helpful than the consistently similar computer. People actually found the "ingratiator" charming: they liked the computer more and enjoyed the interaction more than those who worked with the computer that was similar from the start.

What was going on? It turns out that people are influenced by what psychologists Eliot Aronson and Daryl Linder call "gain" effects. The central idea is that when people first get a small reward and then receive a large reward, they feel even better than when they receive a consistently large reward. Researchers have also observed these gain effects in investment behavior (a stock that suddenly produces a large return is valued more than a stock that consistently provides large returns), salaries (a small raise followed by a large raise feels like more recognition than a consistent raise, even if one receives a higher net amount with the consistent raise), and even dating (people like someone who becomes more interested in them over time more than someone who is interested from the start).

These results also explain a great mystery in social life: despite the strength of similarity-attraction, why do opposites seem to frequently attract? The answer is that opposites attract when, over time, they change to become more similar to each other. Everyone has seen "total opposites" get married and have a happy life together. However, that happiness comes only when the two people change to become more similar to each other, eventually finishing each other's sentences just as married couples who started out similar do.

Becoming similar generates even more positive feelings than consistent similarity-attraction because people perceive it as unspoken praise. That is, when people change to match you, they are implicitly saying, "I realize that you do things the right way and I've been doing things the wrong way. I'm so sure that you are right that I am going to change to be like you." In a world of fragile egos, what could feel better than having people adapt themselves to become more like you? Imitation truly is the sincerest form of flattery.

With an understanding of personality matching, you can easily attain successful relationships: when you interact with someone with a similar personality, just be yourself, and when you interact with someone with a dissimilar personality, try to become more similar in your speech and behavior.

If you find it difficult under the intensity of face-to-face interaction to present a personality different from your own, memos, e-mails, and other written communication provide an easy opportunity to change your language style without having to worry about the range of other personality markers. This is the same technique I described earlier in the chapter that allowed the computers in the eBay and Desert Survival Situation studies to clearly display different personalities to participants. For example, if your correspondent is introverted and you normally write with a lot of adjectives and adverbs, cut one or two from each sentence. If your e-mails generally focus on your own perspective and you're writing to a friendly person, mention your interest in or concern for your correspondent a couple of times. Don't just throw in a single, gratuitous sentence to placate your interaction partner: this type of confusing inconsistency annoys every reader.

- There are four fundamental personality types: extrovert, introvert, critic, and sidekick. You can recognize the four types from language choices, voices, postures, gestures, attitudes, and behaviors. Each type captures a clear set of orientations to other people and the environment.

- Birds of a feather flock together. When you work with people with personalities similar to your own, you will like and trust them more, think they are more intelligent, and even buy more from them.

- Clear personalities are better than ambiguous personalities, even if they do not match that of the person with whom you are interacting.

- Imitation is flattery (or birds that become of a feather really flock together). When you work with people with a dissimilar personality to your own, change to become like them gradually and uniformly. This can be even more effective than consistent similarity.

CHAPTER 3

Teams and Team Building

Whether to avert disaster or reward success, team-building exercises have become an increasingly common part of the business landscape. Although business gurus have touted literally thousands of exercises as key to turning a group into a team, a day of team building almost always follows the same four-part scheme. First comes the warm-up exercise, designed to break people out of their work roles. For example, each person is asked to invent a unique movement, such as pirouetting in place, jumping like a monkey, or waving like the queen of England; the next person has to enact the previous person's movements followed by her or his own. Next is the trust exercise: people demonstrate their faith in their teammates by closing their eyes, folding their arms over their chest, and falling backward shouting "I trust you" while teammates catch them before they hit the ground. Third is the "each of us can succeed only if we all succeed" exercise: for example, the team is given a diverse set of objects and must build a bridge that everyone can get across; if anyone falls off the bridge, the whole team must start over. Finally, there is the "thrilling reward" exercise: hurtling down a raging river in a raft, bonding through terror, is a classic of the genre.

After working as a consultant with a group for an extended period of time, I take it as a genuine compliment when they invite me to team-building events. While delighted by the kind sentiment behind the invitation, I find the actual events a misery. My poor memory, exacerbated by my chronic lack of sleep, makes me seem "unengaged" during the icebreaker. Notorious for my klutziness, I fall backward enough unintentionally that doing it on purpose seems tragic rather than trust building. With my atrocious construction skills, team members have told me that standing still and trying not to touch anything is the best way for me to help finish the bridge. And my weak stomach makes white-water rafting an iffy proposition. Worst of all, I worry that being a somewhat unwilling and less-than-delighted participant in team-building exercises will cement my position as an outsider.

Despite my concerns, I rarely see any difference in how the group treats me before and after our team-building trip (although the trips do significantly increase my store of harrowing memories and nicknames). This experience is by no means unique to a particular set of team-building exercises: time and again, within a few days after a retreat, I would see the team exhibiting the same relationships and the same quality of performance as it had before. I worried that my distaste for team building was blinding me to its benefits, but when I talked to other team members, even people who loved the change of pace admitted that they did not see or feel any lasting effects on the group.

So what goes wrong with team building? Two possibilities come to mind:

1. True teams are an important and effective way to work together, but traditional team building doesn't result in teams actually being built.

2. The importance of teams is overrated: that sense of "one for all and all for one" is nothing more than a nice feeling with no concrete benefits such as improved productivity.

To answer the question of whether the traditional approach to team building is effective, my research team at Stanford (and it really is a team!) came up with an experiment. My first idea was to randomly select groups from within a company to either do team-building exercises or stay at the office as usual. I would then look for differences between the groups in areas such as productivity and cooperation. Thinking about it a little more, though, I realized that the results from such an experiment would be biased. The groups that did not get to frolic in the woods might suspect that management had some hidden motivation for excluding them, which could harm morale and skew the results. Also, each group builds relationships in its own way, so unless we studied an enormous number of people, I wouldn't be able to compare the teams in the two conditions.

I decided that the best way to avoid the psychological biases that come with recruiting and the heterogeneity between naturally occurring teams was to build a team made up of people and computers. If teams are as powerful as people have been led to believe (and because people socially interact with a computer the same way they do with a person), we should be able to see the benefits of teams even when one of the team members is a computer.

I was reluctant to have a computer fall into a person's arms or to strap one to a raft. So I turned to the social sciences, rather than professional team-building consultants, for effective strategies.

How to Create a Team

Ironically, while companies focused on team-building exercises spend enormous sums of money in the belief that team formation requires special locations and tremendous effort, social scientists have found ways to reliably create teams out of randomly selected people in a very short time. What are the secret ingredients in the social scientists' "special sauce"? *Identification* and *interdependence*.

The idea behind *identification* is that personality similarity is not the only means of bonding. Similarity-attraction, probably the most well-established principle in all of social psychology, dictates that you will feel bonded with anyone else who is similar to you on almost any dimension. Thus, when team members share one or more characteristics that they generally do not share with those outside the group, it cements the group together.

The core question in creating group identification is how similar the members must be to become a team. To examine this question, I started with the most extreme form of similarity: interacting with oneself! Would someone who is virtually an identical twin represent the ultimate similarity or an eerie and repulsive doppelgänger?

Experiment:

How Do I Love Me? Let Me Count the Ways

My idea was to compare interacting with a virtual twin versus interacting with someone else. Participants had to feel that they were really interacting with themselves, rather than simply looking in a mirror, and the situation had to generate strong reactions so we could observe differences. People would perform tasks and then be evaluated by a face: either their own or that of someone else. I needed the evaluations to feel accurate so that people would not think that they were "lying" to themselves. But because evaluations generate very powerful responses, I also had to ensure that everyone received identical feedback.

To address these constraints, my Ph.D. students Eun-Young Kim and Eun-Ju Lee (now a professor at Seoul National University) and I videotaped all the participants reading a list of positive and negative evaluations, such as "You did a great job," "That was very insightful," "Your performance was poor," and "That will lead you nowhere." About three weeks later, we had the same participants come back to the laboratory to play several rounds of Twenty Questions with a computer (the same task used in the experiments in the introduction and chapter 1). In this version of the game, we had people sit down in front of a computer that was "thinking" of an animal. They had to guess the animal by typing in "yes" or "no" questions such as, "Is it warm-blooded?" or "Does it have four legs?" The computer would then display a "yes" or a "no" in response.

After every few questions, the computer would play a video evaluation of the questions the player had just asked. Half the participants saw a video of themselves speaking. The other half saw a video of someone else of the same gender, race, and age. After the evaluation, another round of questions would begin (not revealing the answers allowed us to give the same feedback to every person). After ten rounds, we gave the participants a paper-based questionnaire that asked about their feelings regarding their "evaluator" and their experience.

➤ Results and Implications

Did people feel a special bond with the evaluator who was "just like themselves"? Absolutely. Based on their answers to the questionnaire, participants who received evaluations from their own face found them more valid and objective than when they came from someone else's face. They also enjoyed receiving the evaluations more. This happened despite the fact that all participants received the same evaluations, none of which were actually based on the participants' performance.

The effects of seeing their own faces also lingered beyond the interaction. Three days after the experiment was over, we followed up with participants, asking them what they remembered about the comments that the computer made. Participants who were evaluated by their own face remembered a greater ratio of positive relative to negative evaluations as compared to participants who saw someone else's face; that is, when interacting with "themselves," they retained the favorable statements and forgot the unfavorable.*

---

* I had a personal experience with the phenomenon of self-preference. My son was a year and a half old, and I had sat him down on the bathroom sink to wash his face. As I removed the dirt, he cooed, "I love you." My heart melted and I said, "I love you too." He looked at me with consternation. "Not you," he said as he pointed at himself in the mirror. "Him!"

Easing up from total similarity, I then wondered about "family resemblances." People sometimes encounter others who look very similar to themselves. Do people feel "family ties" even without genetic linkage? Studying this question would obviously be extremely difficult with human confederates. Fortuitously, two communication professors at Stanford University, Jeremy Bailenson and Shanto Iyengar, and their colleagues came up with an ingenious way to explore this question with computers in the context of the 2004 presidential election.

By one week before the election between George W. Bush and John Kerry, the American public knew the candidates' faces extremely well. Thus, it was very natural for Bailenson et al. to show people pictures of the two candidates and ask how likely they were to support one candidate over the other. The participants in the study constituted a representative sample of the United States, but Bailenson et al. focused on those people who might actually change their minds: independents and people with weak party affiliation.

Bailenson and colleagues had the clever idea to alter the photographs of Kerry and Bush. In the experiment, they presented each participant with two morphed pictures: one picture was one candidate's face merged with the participant's own face, and the second was the other candidate's face merged with a randomly selected person's face. They then asked which candidate the participant would vote for. The degree of morphing was 80 percent candidate and 20 percent participant for half of the participants and 60 percent candidate and 40 percent participant for the other half.

Although people had extraordinary amounts of exposure to the faces of the candidates through media coverage, only 3 percent of participants noticed anything unusual about the morphed picture. Even these particularly observant participants thought that the photos were "touched up"; not a single person realized that her or his own face was blended with the candidate's!

The effects of the face morphing were remarkable and arguably terrifying for the political process (but a great example of the power of

similarity-attraction!). Those people who were morphed with Kerry supported him 47 percent to 41 percent (12 percent undecided), while participants who were morphed with Bush supported him 53 percent to 38 percent (9 percent undecided). In any race as close as the 2004 election, the use of this type of software in targeted ads to individuals (readily possible through Web sites such as Facebook) could sway an election. A follow-up study by Bailenson and colleagues with unknown candidates found even stronger effects of face morphing.

Beyond the world of political advertising, videoconference or videophone systems present opportunities to leverage this morphing technique. For example, as your face appears on the screen, software could morph it with the person you are speaking with. This would create feelings of similarity-attraction and make the conference go more smoothly than it might in person. A multiway videoconference introduces even more opportunities: your face could be morphed differently for each person's monitor.

You can also see the power of similarity in more mundane situations. For example, if you scan the people at the average restaurant table, you will find they will look more like family than like strangers. Similarly, people in a small business frequently look like they are relatives even when the business is not family owned because hiring is often guided by initial impressions, which in turn are guided by perceived similarity.

Beyond virtually identical appearance, more moderate indicators of similarity are also surprisingly effective at making people feel bonded to each other. These characteristics become identifiers of the group, qualities that define and unite them. Thus, age ("Gray Panthers"), ethnicity ("Sons of Italy"), appearance ("The Redheaded League"), as well as other demographic characteristics can unite teams. Bonds can also come from matching interests, such as a shared love for flower arranging, similar knowledge of *Star Wars* trivia, or a common fascination with model trains. The more seemingly coincidental and unusual the similarity, the better: "Wow, we're both left-handed" is good (around 12 percent of the U.S. population); "Wow, we've both been hit by lightning" is even better (approximately 370 cases per year in the United States, 82 percent of which involve men). Finding unusual similarities is remarkably easy: people drastically underestimate the likelihood of coincidences.

The obsession with similarity applies to even the most trivial matches. For example, the world's leading authority on persuasion, social psychologist Robert Cialdini, demonstrated that when people find a lost wallet and the owner's name is similar to their own, they are more likely to return the wallet to lost-and-found than when the owner's name is dissimilar. Researchers have also shown that people are more likely to marry when their first names are similar.

In many cases, similarities within a group are neither clear nor obvious. When this happens, you must consciously manipulate team identity by identifying and then highlighting a shared quality. In a college football stadium, for example, people in the stands are bound by only a single characteristic: the university for which they root. Nonetheless, students and alumni feel perfectly comfortable screaming, "*We* are number one" or "*We* won," even though they had nothing to do with the team's success: they didn't play, they didn't coach, they probably don't know the players personally, and their individual contributions to cheering were insignificant. Despite the obvious divide between fans and football players, "we" can all bask in the success of what "we" achieved.* Similarly, the custodial staff of my dormitory on campus (I am a "dorm dad") takes special pride in the students that live in "their building."

Of course, the greater the number of identifying characteristics that the group can find, the more powerful the identification and the resultant team bonding. Organized sports leagues understand this: teams offer fans the opportunity to wear unique colors (Notre Dame's gold and blue, the silver and black of the Oakland Raiders); laugh at a unique mascot (the "Philly Phanatic" of the Philadelphia Phillies, the Stanford "Tree"); perform unique gestures (the "Tomahawk chop" of the Atlanta Braves, shaking the cowbells for the Sacramento Kings); sing unique fight songs (the "Locomotive" at Princeton, created in 1880 as the first college cheer, "Fight On" for USC); and wear unique apparel (the Green Bay Packers' "Cheeseheads").

Rather than identifying or intentionally creating similarities, people who want to form teams can also encourage and highlight similarities that naturally emerge as they work together. A series of studies by Stanford design researcher Ade Mabogunje and professor Larry Leifer demonstrate that groups bond more when they create words to describe new ideas and then use those neologisms together. For example, they found that design teams work more effectively when they select special names for different versions of a part rather than refer to them by number, as in "the clam scooper" versus "v2.3." Similarly, my lab group has a set of "inside jokes" that relate to our shared experiences—a single phrase such as "and the sequel, 'Yes I

* Is the use of "we" a result of being caught up in the moment, some light-headedness from screaming too hard? No. In a classic field study, Robert Cialdini and colleagues found that approximately 1.5 times as many students wore clothes that identified their university–Arizona State, Louisiana State, Ohio State, Notre Dame, Michigan, Pittsburgh, and Southern California—to class on the Monday after a victory than on the Monday after a defeat. When the football team wins, students find a way to highlight their similarity to the team (demonstrating their bond through what they wear), and the signs of identification last for days.

Can!'" will cause us all to burst into laughter while outsiders look on baffled.*

You can leverage existing similarities, no matter how small or seemingly irrelevant, to build a team. But in most cases, a collaborating group does not have the luxury of systematically choosing its members based on small similarities or the time to let similarities emerge naturally. Fortunately, you can also arbitrarily create bases for identification; these shared traits can have all the power of genuine similarities.

The infamous "color wars" of summer camps exemplify the power of arbitrary labels. In a traditional color war, counselors randomly divide the camp into two teams—for example, the green team and the orange team—that compete in a variety of activities. Almost immediately, green team members notice that people on their team are better, stronger, smarter, more enthusiastic, and more attractive than the orange team members. Seemingly defying the laws of logic, orange team members simultaneously notice that their members are better, stronger, smarter, more enthusiastic, and more attractive than green team members.

The power of arbitrarily assigned similarities extends to adults as well. In one of the most important studies in the field, conducted by social psychologist David Wilder, experimenters first showed participants two badges, each with a different team name. They then gave participants one of the badges and sent them to a room that had a sign on the door and a banner hanging on the wall with their group's name. Participants also saw the other participants go into one of the two rooms, depending on which badge they had. Once in the room, par-

* My lab group was having a karaoke party, and I had just sung dreadfully. As the team laughed at me, I told them, "You'll regret this treatment when you hear my new album, 'I Can't Sing Any Worse.'" At which point, someone retorted, "I'm waiting for the sequel, 'Yes I Can!'" From then onward, whenever anyone on the team says something like, "I'll never do that again," someone else will say, "and the sequel, 'Yes I Can!'"

ticipants sat at desks with large dividers that prevented them from seeing the other people in their room.

Participants were told to imagine that they were consultants giving a company advice about whether an employee was guilty of corporate espionage. If they thought he was guilty, they were to decide on the appropriate punishment among six options of increasing severity: 1) notify his supervisor to keep a closer eye on him in the future; 2) place a letter of reprimand in his personnel file; 3) reprimand him and transfer him to another office in the company; 4) demote him to a lower rank and transfer him; 5) dismiss him from the company; and 6) dismiss him and initiate a lawsuit for damages to the company.

Participants were then presented with a three-page dossier that described the case in a way that made it sound as though the employee was guilty and clearly merited one of the two most severe forms of punishment. After reading the dossier, participants were given written comments that appeared to come from three other people, all of whom recommended one of the two weakest forms of punishment and provided two reasons for their judgment. Half of the participants in each room were told that the recommendations came from their own group. The other half were told that the input came from the other group. After receiving the input, participants were asked to provide their own judgment about guilt or innocence, the appropriate punishment, and the reasons for their decision. They were also asked to assess the quality of the written comments they had received and then tested on how many comments they remembered.

These extraordinarily minimal efforts toward establishing team identification—a label and a specific location—had a strong effect on people's judgments. When participants' own teammates (ostensibly) recommended a slap on the wrist, the participants agreed with the overly mild punishment; when the other team recommended a slap on the wrist, the participants rejected the advice and urged one of the maximum punishments. Participants also thought much harder about the information if they (ostensibly) received it from their own team, remembering more of what they had read in the recommendations. While the teams were arbitrarily formed minutes prior to the exercise and designated with only a badge and a sign, these symbols strongly influenced participants' judgments. The message is clear: create a shared identifier and you have the makings of a more committed team.

While the above demonstrates that people consider their own team to be a more valuable source of information than a different team, how do team members react to someone who is not a member of any team at all? The experimenters addressed this question by adding a condition in which they told participants that they were simply receiving "other people's opinions" without any mention of what team they came from. Demonstrating the power of teams to bond and repel, the unaffiliated comments were midway in affecting participants' decisions about the right punishment and their memory for the comments.

Along with identification, the other key factor for creating a sense of team is establishing *interdependence*. To create interdependence, research suggests that team members must share two beliefs. First, they must feel that achieving the group's goals will also serve their own personal goals. Second, team members must believe that their efforts and the efforts of other team members are integral to the success of the team.

Perhaps the most important research on how interdependence builds teams is one of the most cited studies in all of social psychology: Muzafer and Carolyn Sherif's 1954 Robbers Cave study. The Sherifs took two groups of boys to different parts of a two-hundred-acre Boy Scouts of America camp and had them perform tasks that required tremendous amounts of cooperation within each group. The groups each had to build a latrine, improve the "swimming hole," and get up a mountain. As the boys realized that they had the same goals as those in their group and had to depend on their group members to achieve those goals, each group quickly became a team and chose a name: the

Rattlers and the Eagles. Because the two groups didn't depend on each other and in fact viewed the other as a competitor performing the same tasks, animosity grew between them. Shared activities, such as meals, would degenerate into brawls.

The researchers eventually realized that if interdependence had created the hostility between the two groups, then interdependence could remedy it. They therefore created crises that required *all* of the boys' help to solve. After working together to get water flowing, jointly using a tug-of-war rope to pull down a dangerous partly felled tree, and rescuing a stuck-in-a-rut truck that was carrying food for the whole camp, the two teams merged into one. Hostilities between the groups disappeared, friendships across group lines developed, and by the end of the trip the boys insisted on riding the same bus home.

While interdependence necessitates having a shared goal, everyone does not need to have the same reasons for wanting to achieve that goal. A recognition that team members must work together to realize the goal, regardless of motive, is sufficient. For example, a VP might tell a marketing team to produce a product specification in six weeks, creating a goal for the team. This leads to interdependence even if different individuals have different reasons for adopting the goal: some team members may be excited about the bonus they will get when they deliver the spec; others may imagine how they will become "stars" of the company when their team hits the deadline; and others may bask in the feelings of camaraderie that come as they struggle to complete the project. Whether financial, strategic, or emotional, as long as individuals hide their differing motivations (so that they do not threaten identification) and all team members explicitly support the shared goal, interdependence will strengthen the group.

Identification and interdependence independently build teams, but these strategies are also intertwined and support each other. The reason identification can lead to interdependence lies in a primitive human impulse identified by Oxford evolutionary biologist Richard Dawkins. By studying seemingly altruistic behavior in animal groups, Dawkins found that in addition to self-preservation, all animals, including humans, are most concerned with supporting those who share their genes in order to ensure that their genes continue on in future generations. Humans innately prefer family members as an evolved response because they have a "selfish gene," as coined by Dawkins.

Because you cannot directly determine how similar a person's DNA is to yours (without a large laboratory), the best you can do is see if they are similar to you in appearance, voice, or any other indicator you can observe. And just as people's brains overextend social rules beyond people to other things that they interact with—computers and other technologies—people overextend the concept of "shared genes" to any type of similarity. That is, our brains have expanded the sensible idea "people who share physical characteristics with me share my genes; therefore, we should support each other" to "people who share *any* characteristic with me deserve my support and will support me." Thus, once you find others similar to you, you are innately driven to help them, and because they also want to preserve their genes, you can expect that they will want to help you. This logic causes mere identification to lead to feelings of interdependence. It also provides another reason why, as discussed in the Personality chapter, people prefer and trust people who have similar personalities to themselves.

In turn, interdependence can lead to identification. Interdependent groups share goals, and these can become the basis for similarities to develop among the group, leading to identification that in turn solidifies the team. When group members adopt the same means to achieve these shared goals, it creates even more similarity to bond them together.

Experiment:

It Doesn't Take Exercise to Build a Team

Now that I had a strong understanding of these two team-formation strategies, identification and interdependence, I could use them to answer the question about whether teams are important. The key was to compare a situation in which people shared identification and felt interdependent with a computer to a situation in which people neither shared identification nor felt interdependent with a computer.

I conducted the experiment with my Ph.D. students B. J. Fogg (now a consulting professor at Stanford) and Youngme Moon (now a professor at Harvard Business School). The first step in creating identification was to have the experimenter continually refer to the computer and person pair as the "Blue Team." We put a blue border around the screen of the computer and placed a sign on top reading BLUE TEAM. We also gave the participants a blue wristband (we used a wristband so participants would see the blue reminder as they typed). In contrast, to discourage a sense of identification, we referred to another set of participants as the "Blue Person working with the Green Computer." Although these participants also had a blue wristband, the computer in their condition had a green border and a sign that read GREEN COMPUTER.

To foster interdependence, we told the "team" participants that the final evaluation of their success would depend on a combination of their work and the work of the computer as compared to other human-computer teams: rivalries between teams are a very strong reminder of interdependence. In contrast, we told participants in the non-team condition that we would compare their performance to other people's; the computer's performance was irrelevant. In this case, the rivalry was *not* one team against another but person against person, with the computer as a bystander.

Now that I had designed a way to foster or hinder team spirit between people and computers, we needed a task that would let this spirit, if it existed, play itself out. To ensure that interdependence would be unambiguous, I opted for a cooperative task. For that to work, I had the computer manifest a level of competence that would make it clear that it could be an effective teammate.

My choice was the Desert Survival Situation, the same classic cooperative task described in the previous chapter. We asked participants to imagine that they and the computer had crashed in the desert with twelve items, including a book called *Plants in the Desert*, a rearview mirror, and a large knife. We then told participants to do the best job they could in ranking the items in importance for survival, with 1 being the most important and 12 being the least important. We told them that to do well they should discuss their answers with their "teammate" or "the computer" before making their final choices.

The interaction started with the participant and the computer sharing their rankings of how important they thought each item was for survival. Unbeknownst to the participants, these rankings were designed so that for half of the items, the computer ranked the item as more important than the participant did, while for the other half of the items, the computer ranked the item as less important than the participant did. This ensured that a lively conversation would ensue and that all participants would have the same initial differences in their rankings between themselves and the computer.

The participant and the computer then discussed each item one at a time, with the participant typing out on the computer her or his assessments of the item and then the computer displaying its comments. The computer was not particularly smart (although the participant did not know that). Specifically, for each item, the computer had only two responses: one version when it ranked the item higher than the participant and another version when it ranked the item lower.

Afterward, participants entered their final rankings. We had participants fill out a paper-and-pencil questionnaire that asked a variety of questions about how they felt about the computer, the interaction, and themselves.

➤ Results and Implications

Would identification and interdependence be so powerful that they could make even a computer a full-fledged "teammate"? I started by examining the participants' feelings about the computer based on the questionnaire results. Consistent with team members seeming similar, "teammates" felt that the computer agreed with them more than did non-teammates, and that their teammate approached the task more similarly to how they themselves approached the task, even though the computer did and said the exact same things relative to the participants' responses in both cases. They also found the computer more intelligent and trustworthy.

Furthermore, as with the study involving badges and group decision making, participants who thought of the computer as a teammate were more persuaded by its suggestions: comparing initial and final rankings, teammates changed their answers significantly more often to conform to the computer's rankings, even though they received identical information from the computer. Finally, despite the fact that the computer did *nothing* in its behavior to show any sense of commitment or belonging, the teammates indicated in the questionnaire that they enjoyed the interaction more, felt more engaged in the task, put more effort into working together, and were more interested in succeeding together.

In sum, by implementing very simple strategies that had nothing to do with the computer's actual behavior, the computer became a member of the person's team, thereby eliciting greater cooperation and effort, greater admiration, more willingness to compromise, and more enjoyment from the activity. Here was a simple machine that participants knew had no notion of commitment to the group, no sense of moral or social obligation, no concern with a negative evaluation, and no sense of self or self-sacrifice. Nonetheless, with the slightest hint of being part of the same team, people not only radically changed their perception of this lifeless box, they actually went out of their way not to let it down. If people are willing to do all this for a computer, just think how much more willing they would be to support their human teammates. Thus, a small but careful investment in fostering identification and interdependence within a group will yield great dividends.

With very minor modifications, these strategies can be used in the workplace. Posting team logos on every team member's desk, colocating the team, and constantly referring to the team by the team name (e.g., "Serengeti" was an engineering team I worked with at General Magic that used *Lion King*–inspired names for all aspects of their projects) rather than its function (e.g., "XP 429 international marketing group") all help to establish identification. Creating situations in which group performance truly determines success encourages interdependence (e.g., if the group hits its deadline, everyone gets a free lunch). Conversely, prominently marking everyone's nameplate with a different color and having team members' desks spread throughout the company's premises undermines identification. Allocating bonuses and praise solely based on individuals' performance rather than a team metric reduces feelings of interdependence.

The social science literature also shows that human teams grounded in identification and interdependence provide the same benefits as human-computer teams. In situations ranging from rapid creation of crisis teams by encouraging "swift trust" (a term coined by business scholars Debra Myerson, Karl Weick, and Rod Kramer) to engineering teams working on an intensive one-year design project (the specialty of Mabogunje and Leifer) to multiyear studies of migrant workers (described by social activist Maurice Jourdane), research shows that people who feel part of a team:

- Act more cooperatively
- Feel happier
- Feel a stronger sense of belonging, control, self-esteem, and meaningful existence
- Make more correct decisions
- Make decisions more rapidly

For example, in his book *The Checklist Manifesto,* writer and surgeon Atul Gawande collects numerous examples of how working as a team is crucial for avoiding failures in organizations. With the sheer volume and complexity of knowledge required for most operations today, he argues, specialists must work together as a single unit to be effective. His findings confirm that simple measures toward fostering a sense of teamwork lead to fewer mistakes and better work experiences. For example, in a study at Kaiser Hospitals in Southern California, when hospital staff's average rating of the teamwork climate rose from "good" to "outstanding," employee satisfaction rose 19 percent and the rate of operating-room nurse turnover fell from 23 percent to 7 percent. More near-errors during surgical procedures were also caught.

Similarly, in contrast to FEMA's top-down approach to responding to Hurricane Katrina, a case study from Harvard's Kennedy School of Government credits Walmart's strong sense of teamwork in its successful contribution to the relief effort. Management gave employees a single, common goal (emphasizing interdependence) to do what they could to help. As a result, Walmart employees came up with and implemented many innovative and effective solutions, including temporary mobile pharmacies and clinics and free check cashing for payroll. And they were able to supply food and water to refugees a day before the government arrived in the city.

Teams can also help new members quickly integrate. As well described by business consultants Jon Katzenbach and Douglas Smith, "newbies" often benefit from joining strong teams because the well-established feelings of interdependence result in the new team member being seen as less of a burden and more of a benefit, even while she or he gets up to speed. Also, because members of a team behave more similarly to each other as compared to mere groups, appropriate behaviors become highly visible, making it easier for the new team member to learn appropriate behaviors.

Why Don't Team-Building Exercises Work?

This brings us back to the question posed at the beginning of the chapter: if teams are so beneficial, why don't team-building exercises result in more effective and cooperative groups? The problem is not with the goal of strengthening teams but that team-building exercises do not focus on identification and interdependence. In fact, they undermine both. For example, the harrowing "trust" falls that require you to muster the confidence that your teammates will catch you don't *create* trust; at best, they only test whether trust already exists. Furthermore, trust does not lead to team formation (it supports neither identification nor interdependence), although identification and interdependence both lead to trust.

The crossing-the-bridge task does involve high levels of interdependence, but other aspects of the task undermine that feeling. As discussed in the Praise and Criticism chapter, people notice negatives such as mistakes and poor judgments much more than positives such as successes (I will talk more about this in the next chapter). In situations of high intensity, this difference is amplified. Thus, as each person haltingly builds the bridge, the group will focus on its difficulties and feel frustrated rather than bond as a team. As a result, once everyone gets back to the office, you are more likely to hear, "Your lack of focus really messed us up on the bridge. You better be a lot more careful with our product report next week," than "I now know that we can depend on each other."

The raft ride does equally little to support team formation. When you next see the people with whom you have shared the death-defying raft ride, you will be reminded of the experience. As will be talked about in the next chapter, this will cause your thalamus and sympathetic nervous system to send a clear message to your brain: "Be afraid. Be very afraid." You can overcome this negative effect with conscious thought, but it does not form a great basis for the typically comfortable interaction among teammates.

It is now clear why traditional team-building exercises do not garner the positive effects of teams: they do not build teams. By failing to create or sustain identification and interdependence, these exercises cannot result in better performance, greater cooperation, stronger commitment, or the other benefits of teams.

How to Successfully Build Teams

Perhaps the most important lesson about team building is that you cannot implement the engines of team formation for a day or a week and then discard them: continual reinforcement of identification and interdependence are integral to the health of teams. Hence, having a team-building retreat is an oxymoron. How then can you make team building a continual activity?

Markers of identification sustain teams best when they increase in quantity and exclusivity. Actively seek and create inside jokes, neologisms, and other opportunities to mark your special bonds, and jealously restrict these shared phrases to your own team. A marvelous example of the effects of diluting linguistic shared identity comes from the ingenious strategy of Stetson Kennedy to subvert the Ku Klux Klan in the 1940s (described in Steven Levitt and Stephen Dubner's book *Freakonomics*). After infiltrating the group, Kennedy discovered that the Klan had a special set of words and phrases that marked membership. For example, asking for "Mr. Ayak" was code for "Are you a Klansman?" their bible was the "Kloran," and the local chapter was called a "Klavern." Understanding how the Klansmen felt bonded by their own secret code, Kennedy had journalist Drew Pearson disseminate, via radio, these markers of identification to millions of people throughout the country. Once everyone had these symbols of team membership, they no longer *were* markers of membership. The effects on the KKK were devastating: meeting attendance and membership applications declined dramatically.

You can also sustain identification by finding a competing team. For this approach to work, the other team should be both highly visible and similar enough to warrant comparison, as in the Robbers Cave study. In companies, an opportunity for this occurs when one team is working on the next version of an existing product or service while another group works on a riskier rethinking of the product. Even though these groups are not technically "in opposition" because they are part of the same company and doing different work, this represents a chance for each team to highlight shared goals within its own team that do not exist (or at least are less evident) in the other team. Thus, a team can note that "they just missed another deadline" ("deadlines are important"); "they never hang out with each other" ("we should socialize more"); and "they won't do anything without their boss's permission" ("it's okay to make decisions on your own"). This also explains the compelling rivalries between public and private schools and between police officers and firefighters. On the other hand, a manager's description of her team at a former company in contrast to her current, presumably superior, team does not generate bonding: it is too distant a comparison for the current team members (as well as a suggestion that disloyalty is acceptable).

Just as people want to feel identity, interdependence, and the benefits of teams with those who are "one of their own," people also want to distance and differentiate themselves from teams that are "too close for comfort." This active rejection can motivate teams as powerfully as the bonds of similarity. Of course, if this strategy gets too extreme, it can lead to active sabotage of one team by another.

The group T-shirt provides an instructive example of a poorly implemented team-building strategy. You can't work in high-tech very long without acquiring a number of free T-shirts sporting group logos and slogans. (In fact, I have so many T-shirts that I now appropriate them for nonapparel uses, such as propping up shaky table legs.) The idea behind these T-shirts is that by reminding you to identify with the group, the shirts will turn your group into a team. Unfortunately, while based on the right principle, the T-shirt ritual in practice more often has the opposite effect. For example, if some types of workers—such as managers, women, and new employees—do not normally wear T-shirts and so don't wear the group T-shirt, the "team" apparel becomes a marker of division rather than a strategy of integration for the group. Instead, put the team name or mascot on clothing appropriate for everyone to wear to the office.

You should gear the design of the clothing toward the group and not to the public at large. Make the identifier large enough so that team members can immediately recognize it, but small enough so that it doesn't feel like a billboard. Rings that commemorate a certain number of years of service at a company are an excellent example of a small and within-group targeted token of identification. Think of these markers as a "secret" handshake.

Also make sure to control when the clothing gets worn. Although some occupations require uniforms, in most workplaces, people ridicule anyone who wears the same clothes every day. Therefore, coordinate and select certain days (e.g., the first Monday of every month) on which *every* team member wears the piece of clothing. Make sure to have extras handy on these days so that if someone forgets, she or he will still be able to share in this marker of team membership. While ensuring ubiquity of the clothes among team members might seem trivial, it is actually very powerful: it symbolizes the importance of both the team and every team member on it. On all other days, no one should wear the team-marked clothing.

Sustaining strong feelings of interdependence is much more difficult than sustaining identification. The problem is that a strong norm exists, at least in the United States, for companies to organize incentives at the individual level with a very large component of compensation, such as raises and bonuses, based on what each person does independent of the overall team's performance. This is a particularly acute problem when rewards are a zero-sum game within a group; that is, when the manager of a group has a fixed sum to distribute and must differentially allocate it to team members. As team members feel that the success of one person leads to the failure of another, they start to realize that they are not as interdependent as they once thought.

Given how unlikely it would be to overturn the notion that incentives must contain an individual as well as a group component, what can you do? Emphasize the intrinsic benefits of cooperation rather than interdependence. That is, instead of making cooperation a means to the end of achieving a shared goal (interdependence), make cooperation the *goal* itself. Change the orientation from, "If everyone doesn't do her or his share, the group will fail," to "It feels great when we *cooperate*." Indeed, there is evidence that people are evolved to enjoy cooperation. In an article that became one of the founding works of sociobiology and sparked my own interest in the field, political scientist Robert Axelrod and evolutionary biologist William D. Hamilton showed, using computer simulations, that groups of animals that cooperate have better prospects of survival than groups with almost any other strategy.

Merging Teams

A common situation that presents a challenge for interdependence strategies is when groups that previously competed suddenly need to work together—for example, after a corporate merger or acquisition. For CEOs, CTOs, and the like, a merger automatically brings interdependence: both companies jointly create profit and loss. For people in lower levels of the organization, however, mergers create significant asymmetries that undermine the sense of a team. For example, if the companies adopt the reimbursement practices of company A, the employees of company A become the teachers, and the employees of company B become the students. This results in *dependence*, not interdependence, between the two sets of employees, which in turn leads to *stronger* identification with the company of origin and *decreases* the identification with the new, unified entity. Furthermore, it creates a situation in which the employees of company A "know everything," leaving the "students" of company B to feel that they are not essential to success.

Two strategies can remedy this dependent relationship. First, you can balance the asymmetry by making the employees of company A dependent on the employees of company B in another, equally important domain, thereby creating interdependence through a division of labor. Second, you can create as many situations as possible in which members of companies A and B must rely on each other to achieve a *joint* solution. Under this approach, decisions and solutions concerning strategic and tactical problems should occur *only* when members of both teams are together, thereby restoring a sense of mutual dependence.

Another challenge for interdependence comes when a member of the team undermines it. For example, if team members prioritize their own personal goals, they might abandon their contribution to the team's project to take on tasks that provide more individual recognition, fully intending to bask in the credit for the team's accomplishments as well. A less pathological form involves taking on the most visible part of a project rather than the part that would most help the team, or taking credit for parts of the project that were actually done by many team members. In addition to inhibiting the team's performance, these "credit stealers" or "show-offs" undermine interdependence.

There are a few strategies that can encourage resistant people to become full-fledged team members. For example, focusing on the upside of team membership can help. Highlight the beneficial aspects of identification and interdependence, such as the great product you will create, the impact that your ideas will have on the rest of the company, or the special camaraderie your team shares.

You can also make working as a team desirable by making the team exclusive. Just as Groucho Marx wouldn't belong to any club that would have him as a member, people tend to want to be members of clubs that are hard to get into. In one of the founding studies of the field of "cognitive dissonance" performed by the great social psychologists Eliot Aronson and Judson Mills, female college students were told that they had to pass a test to become a member of a group. In a technique that seems quaint in the twenty-first century, the female students in the *severe* initiation condition had to read twelve obscene words and two "vivid descriptions of sexual activity" from contemporary novels. The *mild* initiation condition required the students to read five words related to sex but that were not obscene. In the third condition, there was no initiation of any kind. All participants then heard a discussion in which the group they were going to join had "one of the most worthless and uninteresting discussions imaginable," according to Aronson and Mills. Consistent with the idea that difficulty of admission makes membership more desirable, the females who had to go through the "torture" of reading obscenities aloud found the group members and the discussions of the group more desirable and interesting than did either of the other two groups.

In the workplace, you probably can't make team members go through hazing or an initiation process, but whenever possible, remind them that they were "selected" rather than assigned and that their accomplishments warrant their inclusion in the group. The more you can say about the rigors of selection and of the number of people who want to be in the group but cannot be, the better.

If these strategies don't work, then it's best to cut your losses and instead try to help fortify the sense of team among the other group members. This requires a more drastic measure—distinguishing the deviant. While singling out a nonperforming team member might seem to threaten team unity and identification, a large number of studies, the first conducted by Émile Durkheim, demonstrate that deviants actually play a key role in establishing strong groups by setting the bounds of appropriate behavior.

Many World War II movies that depict life in the military demonstrate how a deviant can strengthen a team. In the barracks, there was frequently one soldier whom the other soldiers saw as a failure. The soldiers would mercilessly tease this deviant and would grumble that they had to pick up the slack from the screw-up. After finding this out, the "brass" would remove the screw-up soldier from the barracks to improve morale. Did the team applaud this effort? Absolutely not. Instead, the other soldiers were furious that their object of ridicule and frustration was gone, and they protested until the soldier was returned. The traditional line was, "He may be a screw-up, but he's *our* screw-up."

Social scientists can predict what would have happened if the higher-ups did not return the screw-up to the barracks: a different soldier would become the new deviant and object of ridicule. This new deviant would have her or his minor deficiencies highlighted. Thus, even teams that do not have an obvious screw-up may create one to strongly establish the norms of the group (just as using another team as a bad example helps strengthen your team).

This principle applies to groups outside the military as well. In my lab, one member was the butt of all of the team's jokes. He was a very nice and intelligent person, but it was as if he had a permanent KICK ME sign on his back. While other members of the team could accidentally let slip an unfortunate sentence or commit a faux pas without any undue attention, the team immediately called him out on his mistakes, albeit affectionately. While the quirks of other team members were ignored, every one of his oddities prompted a shared laugh. When he said, "I have an idea," someone else would say, "Tell us before it dies of loneliness." Even when he overcame his perceived deficiencies, the success was termed a "miracle." Nonetheless, we were all devastated when this special person graduated; I'll be watching for someone to fill his role while searching for alternative strategies to bond the team.

So don't paper over the differences between a lacking team member and the rest of the team. Instead, use the deviant to make clear what binds the team together.

Groupthink and the Suppression of Diverse Opinions

Sometimes the benefits of a team can actually become a detriment. In most cases, teams reach much better decisions than simply the average of each member's opinion. Clear advantages come with hearing a variety of viewpoints and deliberating with an understanding that disagreement serves the team's goals. However, a demand for uniformity can come from the perception that disunity undermines similarity (identification) or that dissension impedes the team from achieving its shared goals (interdependence)—picture a team of horses pulling in different directions. When people think that all decisions must be unanimous, they suppress alternative opinions and the team comes to decisions that all parties accept but that fail to leverage the benefits of multiple viewpoints.

Yale psychologist and pioneer in the study of social dynamics Irving Janis identified the problem of *too* much agreement by coining the term "groupthink" more than thirty years ago after studying the Kennedy administration's decisions on the Bay of Pigs and intervention in Vietnam. Janis argued that some of "the best and the brightest" people in the country made terrible decisions not out of ignorance but out of a desire to be agreeable. NASA's investigation into the cause of the space shuttle *Columbia* disaster reveals an example of the problems with groupthink. While some of the engineers connected to the mission were satisfied that falling foam during the shuttle's launch did not endanger the craft, other engineers consulted were concerned that the material could potentially cause catastrophic damage. These minority views got stifled and were not passed along to the important decision makers, and as a result no action was taken. Although the people making these decisions were literally rocket scientists, they fell into the trap of focusing on agreement rather than finding the right answer. Organizations or teams that have very strong norms of "loyalty equals agreement" are particularly susceptible to pressures for unanimity.

You can moderate a group's compulsion to uniformity with Abraham Lincoln's approach to building his cabinet, as described in Doris Kearns Goodwin's *Team of Rivals*. Lincoln appointed people who had run against him and who didn't necessarily like or respect him but whom he nonetheless believed would make important contributions. Similarly, President Obama said in a press conference announcing his cabinet appointments that he was combating groupthink by gathering "strong personalities and strong opinions," including his past rival Hillary Clinton and Robert Gates, director of the CIA under President George H. W. Bush. By bringing together people with very different views and highlighting that you have brought them together specifically to benefit from their diverse knowledge, you combat the idea that being a team means agreeing with each other. Remind the team members that they are bonded by working on the same problem and by striving to understand each other's points of view to find the right answer. Conversely, de-emphasize that the group is a decision-making body, which implies that they must reach consensus and urges people to bury dissent in a misguided attempt to accomplish the team's stated goal.

You might think that a desire for uniformity would lead teams to make moderate, if not optimal, decisions. Research initiated by MIT master's student James Stoner in 1961 and later confirmed in hundreds of studies reveals the opposite: teams committed to unanimity tend to polarize rather than moderate their decisions. In most situations, groups become overly optimistic, leading to a "risky shift." That is, groups tend to select an option with a large upside, even if the downside is larger and more probable, because they focus on the positive.

This shift to the positive occurs, as noted in the Praise and Criticism chapter, because people like positive commentators more than negative ones. When people feel identification with a group, the desire to be liked and accepted can overcome their desire for accuracy or perceived intelligence (both of which would involve more negative comments). The preponderance of a positive viewpoint makes the optimistic decision seem average, leading groups to converge on a risky or optimistic outcome. Thus, when a group is choosing among a number of choices, you should be suspicious when the group moves to steadily more risky options.

You should also be mindful of how you refer to the group's opinion versus your individual opinion. In general, teams benefit from references to "we," "us," and the group name. When supporting another member's dissenting idea, you might say, "We really benefit as a team from each of our viewpoints." But fight the tendency to speak for the group ("I'm glad to see we're all on the same page") until it actually reaches a decision—this can discourage dissent.

Very few teams are "total institutions," a term coined by sociologist Erving Goffman to refer to groups that interact only among themselves. More typically, individuals in a team may also work on projects with other teams or collaborate with people outside their company altogether. In these cases, outsiders such as clients or suppliers sometimes attribute characteristics to the team member that reflect stereotypes of the team.

I once worked with an engineering group that had gained a reputation as impulsive (in my opinion, simply because the group believed in rapid prototyping). When I went to a meeting in a different division with a member of that team, every one of his comments was either ignored or greeted with, "You haven't thought through the options." I realized that people's perceptions of the guy were biased and that it was important to distinguish him from the reputation of his team. As an "independent" observer (one of the nice things about being a consultant), I made a point of constantly using his name and never mentioning his team. Shaking the group stereotype was difficult, but eventually he became as influential as he deserved to be.

Membership in a group can have negative consequences even when the group has an excellent reputation. A striking example of this effect comes from my consulting work with Dell Computers. Dell asked me to develop a pictorial agent on a Web site used to help people customize their computer. This agent, represented by a photograph and a word balloon, could give people advice on the best computer components for them; the agent would even cross-sell and up-sell. So, for example, if a buyer selected a fancy audio board, this agent might recommend high-end speakers. The goal was not only to engage and help customers but to improve profit by suggesting items that the purchaser might not have thought of. Just as every order-taker at McDonald's says, "Do you want fries with that?" the agent would say, "Do you need a printer?"

In selecting the perfect character, we started with photos of more than 250 male and female models. We eventually invited twenty of our favorites for a photo shoot and circulated the photos to hundreds of people in the company and numerous focus groups. By the end, I had found the perfect guy: he was very attractive and seemed both friendly and intelligent. He filled everyone with a sense of trust and the feeling that he had their interests at heart. The enormous expense of the search, I believed, had clearly been worth it.

Dell placed the agent on the Web page with the scripts that my team and I had carefully written, and the testing began. The results were puzzling. Despite all of our work, people just didn't seem to trust our agent. It wasn't just our questionnaires that suggested something was wrong; typical comments in interviews included, "This guy doesn't really care about me; all he cares about is selling," and "I bet they put him there just to make money."

We went through our usual design audit and made sure that we had incorporated all of the social principles we had developed over the years. Everything was there: the agent praised the customer, he had all the markers of honesty and sincerity, and his facial expressions perfectly matched the words he was saying. Finally, someone on the design team shouted, "I've got it—it's the damn shirt!" We had outfitted our agent in a shirt of Dell blue with the Dell logo of white letters displayed prominently on the front. The shirt revealed the agent's true loyalties.

We brought the model back into the studio and reshot all the photos with him in a comfortable red shirt without any logo. The results (obtained from questionnaires) were dramatic: people found the agent thoroughly credible and trustworthy. Participants were much more likely to take the agent's recommendation, and they also felt very strong positive feelings toward him. Sales and profit margins increased. While many Web sites worry that asking too many profile questions will annoy customers, the users who worked with this agent were delighted to provide more and more information, knowing that (as participants indicated when they agreed with the following statements on the questionnaire), "he really wants to understand me" and "he'll put the information to good use." My favorite moment of the study came when an experienced corporate IT person spontaneously said, "I really like that guy. He's on my side, not on Dell's side."

Perhaps the most obvious example of leveraging this phenomenon outside of the world of computing comes from the automobile industry. Car salespeople will often say something like, "You are a very nice couple. I'm not going to let you buy this car because it's not right for you. It's true that we make the most money on selling this one, but I just can't do that to you." Savvy salespeople imply through this that they are abandoning their company team and becoming a team with the customers. They even leverage the trick of scapegoating their natural team, following the lesson of Microsoft's Clippy (as I described in the Introduction): "My boss is going to kill me, but I'm going to challenge him to get you the car at this price. He's obsessed with every penny, but I'm committed to making this work." Salespeople will also use all of the other strategies of "pseudo-gemeinschaft" (a concept coined by Robert Merton and made popular by my Ph.D. advisor, sociologist James Beniger)—the false impression that someone is doing something "just for you." Of course, the best salesmen don't limit themselves to leveraging the team mentality: they use all of the strategies discussed in this book.

- Teams are created by identification (characteristics shared among team members) and interdependence (a feeling that the team is working toward a shared goal). Traditional team-building exercises don't build teams because they support neither identification nor interdependence.

- Establish identification by sharing characteristics. These characteristics might exist beforehand, but you can also create identification with something as simple as a shirt (if managed properly).

- Establish interdependence by emphasizing the importance of cooperation and by recognizing all team members' contributions.

- Identification encourages interdependence, and interdependence encourages identification.

- When part of a team, people will feel happier, act more cooperatively, and make better decisions.

- Watch out for the pathologies of teams such as groupthink, the tendency to select extreme options, and being stereotyped by people outside the group.

CHAPTER 4

Emotion

One day back when my son, Matthew, was two years old, he was gobbling down his food. "You shouldn't eat so fast," I told him. "Throwing food in your mouth without tasting it insults the cook." Matthew stopped eating and gave me a defensive look. "How come you don't yell at Honey when she eats fast?" he said. "Dogs don't know that chefs have feelings," I replied.

That evening, Matthew and I were taking a walk when we passed a bulldog standing in front of a house in our neighborhood. Matthew turned to me and said, "Dad, look at that ugly dog." "Don't say things like that," I chided him. "You'll make the dog feel bad."

"So feelings about eating don't count for dogs but feelings about looks do?"

Here was a moment when my decades of teaching and consulting should have shined through as I guided my son through the complexities of life. Instead, I sighed and used a parenting trick that I knew always worked: "Let's get some ice cream," I said, ". . . and we'll eat it slowly."

While a child (or a perplexed father) may struggle with the emotion of dogs, all people struggle with understanding and managing their own emotions as well as those of other people. Confusion surrounding the proper role for emotions particularly plagues people on the job, as messages conflict about which emotions are appropriate when. On the one hand, classical organization theory prescribes that emotion is the enemy of order and efficiency. Hierarchies, operating policies, and training regimens establish rules and structures precisely to restrict the effect of any given individual's passions and emotions. Managers are supposed to discourage overt displays of emotion: "There's no crying in baseball [or in any other job]!" On the other hand, in my work as a consultant, I have seen the crucial role emotions can play in organizational success. If employees have no emotion about their company, they simply do exactly what is specified, no more and no less. It is only when employees are driven by their commitment to an organization to do the "right thing" rather than the "formally specified thing" that companies can truly thrive. Indeed, in *Fortune* magazine's list of "Best Companies to Work For," the majority of the companies are praised for their corporate culture and support for workers rather than for their salaries and layoff policies. For example, SAS (ranked first in 2010) prides itself on a culture based on "trust between our employees and the company." One development tester describes the company as so supportive that "it feels like a family—almost an extension of home." Robert W. Baird & Co. requires that applicants pass its "no asshole" rule: bullies will not be tolerated.

What managers explicitly say about emotion confuses things even more. I once worked with a manager who would often espouse, "Fun is job one!" But whenever he ducked his head in to see what the team was up to and said something like, "You guys certainly seem to be having a good time," the team would take it as criticism. Everyone enjoys the banter of a witty coworker, but colleagues also warn against jokes that can offend and create a hostile work environment. People often talk about the importance of passion, but when someone gets "carried away" and starts yelling during the conference call, she or he is admonished to "keep it together" and "not take things so seriously."

Emotions seem a particularly complicated domain for which to derive a small but powerful set of social rules. First, there is the sheer number of emotions: English has almost one thousand words that describe emotions, each with its own nuances. How can you summarize all of these different feelings? Second, emotional responses seem to differ dramatically from person to person: some people are extremely expressive and literally beam with happiness from every pore of their body, while others might just crack a tiny smile or show a little twinkle in their eyes when experiencing the same amount of delight. Fortunately, our research and the social science literature demonstrate that emotions are much simpler than they appear. Specifically, although emotion is intimately tied to everything from the most primitive parts of your brain to your highest-thinking processes, almost everything that you need to know emerges from two fundamental concepts: valence and arousal. With this framework, you can answer the following questions:

- Which emotions benefit which situations?
- How can you leverage and manage your emotions?
- How can you leverage and manage other people's emotions?

People frequently find it difficult to describe how they feel; there seem to be so many shades of emotional feeling. Despite this apparent complexity, social scientists, beginning with the work of psychologists Charles Osgood, George Suci, Percy Tannenbaum, and Peter Lang, have found a straightforward way to categorize emotions. To expose my Ph.D. students to this scheme, I have developed a rite of passage. I tell each student that I want them to design a questionnaire that will be used to assess participants' emotions about an interface. I tell the students that they can use as many adjectives as they need to explore the complexity of emotion.

I have done this dozens of times and, at first glance, participants' responses seem remarkably varied and subtle. However, applying statistical techniques to the adjectives reveals a startling pattern. Despite the ingenuity of my students, their instincts about human behavior, and the length of their lists, all of the adjectives can be summarized very simply by answering only *two* questions! This occurs even when participants slowly and carefully ponder all aspects of their emotional life.

What Is Emotion?

The key to this approach is the finding that people in all cultures face every situation by asking themselves, both consciously and unconsciously, two questions:

1. How well am I meeting my goals?

2. Should I do something about my goals?

We also can frame these questions in terms of feelings:

1. How happy am I?

2. How excited am I?

The first question, called the valence question, reveals whether you believe that you are meeting your goals—thereby feeling happy—or whether you are failing to meet your goals—thereby feeling sad. The second question, called the arousal question, is like a volume knob on your responses: are you vigorously trying to meet your goals—excitement—or are you letting the situation play itself out—calm?

Looked at another way, valence is the judging side of people: where am I in terms of where I want to be? Arousal, conversely, is the doing side: am I ready to act? Thus, emotion is an extremely concise and efficient way to link people's goals, their current situation, and their attitudes and behaviors.

Emotions are not just about yourself; slightly modified versions of the valence and arousal questions can help you understand the feelings and goals of others as well. That is, to determine others' emotions, simply ask: "How happy or sad are they?" and "How excited or calm are they?"

How can valence and arousal summarize all emotions? Let's start with the extremes. If you can say that you are clearly meeting your goals (very happy) and are actively responding to that feeling (very excited), it is called "ecstasy." "Serenity" or "nirvana" describes the feeling of extreme happiness, like ecstasy, but combined with feeling very subdued. Like serene people, "despairing" people are very subdued, simply wanting to stay in bed, but they feel strongly negative instead of positive. Finally, if you feel far from achieving your goals (very sad) and are striving to change (very excited), the emotion is "rage," such an active state that you may find your body actually shaking.

If one thinks of valence and arousal on a graph, with valence as the horizontal axis—ranging from very negative to very positive as one goes left to right—and arousal as the vertical axis—ranging from very calm at the bottom to very excited at the top—ecstasy, serenity, despair, and rage would each fall on one of the four corners of the graph.

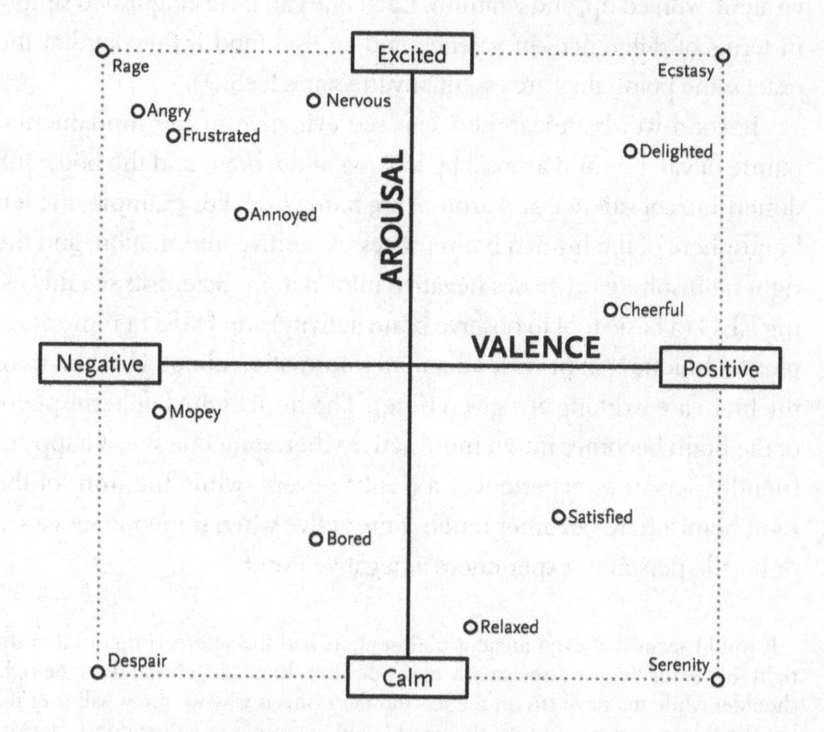

It is also easy to plot and describe more moderate emotions by determining their extent of valence and arousal. "Cheerful," for example, is quite happy and a little more excited than neutral (toward the right edge of the graph, slightly above the horizontal axis). "Mopey," conversely, is quite sad and a little more calm than neutral. "Relaxed" is just a little bit happy and quite calm, while "nervous" is a little bit unhappy but quite excited.

This means that to understand your own or other people's emotions, you don't have to be a poetic soul making incredibly subtle distinctions. Instead, you can discern even nuanced differences in emotion merely by identifying levels of valence and arousal. For example, people recognize many different types of anger: annoyed, apoplectic, appalled, boiling, cross, disgusted, enraged, frustrated, fuming, furious, hostile, in a huff, in a stew, incensed, indignant, inflamed, infuriated, irate, irked, irritated, livid, mad, miffed, outraged, piqued, rageful, up in arms, vexed, virulent, worked up, and wrathful. Each one can be distinguished simply in terms of differences in valence and arousal (and if they land at the exact same point, they are essentially the same feeling).

Beyond words, researchers can see evidence of the fundamental nature of valence and arousal by looking at the brain and the body: the dimensions of valence and arousal are hardwired. For example, the left hemisphere of the human brain processes positive information, and the right hemisphere processes negative information. Scientists see this using EEG (a basic tool to observe brain activity) and fMRI (a more complex technique that provides detailed information about which parts of the brain are working at a given time). The front of the left hemisphere of the brain becomes much more active when someone sees a happy or friendly person or experiences a positive event, while the front of the right hemisphere becomes much more active when someone sees a sad or hostile person or experiences a negative event.*

---

* It would seem that even ancient philosophers had the (correct) notion that the right side of the body is associated with the positive. In art, angels stand on the right shoulder while the devil sits on the left shoulder (this is why we throw salt over the left shoulder when we spill it); the word "right" referring to a direction is derived from the Old English word for "correct," and "left" from the Old English word for "foolish"; and the Latin word "dexter," meaning "on the right side," came to mean "auspicious," while "sinister," from the Latin "on the left side," led to words meaning "ominous," "bad," and "wicked."

Arousal is separated not by sides of the brain but by specialized parts of the nervous system. The sympathetic nervous system (originating in the spinal cord) controls excitement and the fight-or-flight response, while the parasympathetic nervous system (originating above the sympathetic nervous system in the medulla and sacral region) controls calm (also known as the rest-and-digest response). When the sympathetic system is more active than the parasympathetic, you are more excited, and when the parasympathetic dominates the sympathetic, you are calmer.

Before messages get sent to the front left or right hemisphere of the brain and the sympathetic or parasympathetic nervous system, various parts of the brain determine what valence and arousal you should be feeling. The brain has a basic system for an immediate response and a slower, higher-level system for more fully assessing valence and arousal. The low-level system includes the most primitive parts of the brain. These make the first and most rudimentary decision as to whether a change in the environment supports or works against your goals (answering the valence question) and whether you should do something about the situation (answering the arousal question). These judgments are very simplistic and frequently referred to as "impulsive"; they may also guide your use of "thin slices," very rapid assessments (as mentioned in the Personality chapter), when judging situations or evaluating others.

Usually, the primitive parts of the brain provide the first interpretation of which emotion is warranted. Once the primitive decisions are made and the primitive responses are executed, the "cognitive system" makes a more complex and intricate judgment. The cognitive system understands language; has a very rich and nuanced view of people, places, and things; recalls both recent and distant events; puts together long chains of cause-and-effect sequences; develops plans and strategies (taking into account multiple factors at a time); and brings cultural norms to bear in interpreting the "emotional meaning" of events. The higher-order parts of the brain integrate the information from the primitive parts of the brain with what is happening within the body and in the environment to make the final decisions concerning the appropriate emotion you should feel and how to manifest that feeling.

I like to think of a woman's typical response to the classic wedding proposal as a beautiful example of how the primitive brain and cognitive system work together. You might think that there is nothing more delightful and romantic than having someone go down on one knee and say, "Will you marry me?" My training as a social scientist makes me see it differently (although I did propose this way myself). If you watch a woman's initial reaction to seeing a man start to get down on one knee, you will see extreme *negative arousal:* a tense body and a face white with fear. This response is due to the woman's primitive brain saying, "He is moving toward the ground. He's either falling, hiding, or getting ready to attack me! I must prepare for something very bad!" So instead of immediately showing joy or surprise, she shows concern about a body hurtling downward! It is only after a moment that the higher-level parts of her brain deduce the symbolic meaning of the gesture, and, assuming that she plans to accept the proposal, her body and face transition to the appropriate joyous state: still very aroused but also very happy. While this transition can happen in less than a second, if the man sees the woman's initial response, he too will exhibit an immediate negative reaction, until the woman's adjusted response snaps him back to his own happy and excited state.

Once the brain determines the appropriate valence and arousal associated with a particular stimulus, it directs a host of bodily activities. People experience a clear and distinct constellation of brain and body markers when feeling an emotion that guides everything from your tapping feet to the feeling of butterflies in your stomach to the crinkle of your brow. That is, the brain, once identifying the emotion being felt, sends a specific set of messages to specific parts of the body in response to that emotion. Thus, one key reason that we can simplify emotions is that all people's brains tend to send the same messages to the same parts of the body for each given emotion. Knowledge of the specific physical reactions associated with feeling different emotions is an extremely powerful tool if you look carefully enough.

When you feel positive valence, the corners of your lips and your cheeks move upward into a smile, your arms and legs spread outward and move smoothly, you sit up straighter, your eyebrows separate, and your voice inflects upward, with sentences sounding slightly like questions. When you feel negative valence, your lips and cheeks turn downward into a frown, your arms and legs tighten inward with jerky movements, your posture slumps, your brow furrows, and your sentences have a falling inflection. The more extreme your happiness or sadness, the more extreme your bodily responses.*

Psychologist Kip Williams and colleagues demonstrated the powerful linkage between people's feelings and their bodies using a social strategy that leads to extremely negative valence: ostracism. While intentional exclusion might seem like a middle-school strategy that people outgrow and become immune to, it is extraordinarily powerful and people of all ages can use the technique. The researchers had people play a video game, "Cyberball," which involved tossing a virtual ball at random among the participant and two other confederates, while in an fMRI machine. Though it was a game, there was no score or evaluation involved (to avoid the effects of competition). For half of the participants, after a few tosses, the two confederates began to exclude the participant and throw the ball only between themselves.

The fMRI revealed a startling outcome: individuals ostracized by the other players exhibited activity in those parts of the brains that indicate physical pain (the anterior cingulate cortex and the right ventral prefrontal cortex). That is, emotions are so powerful that they can make you physically ache. There were also effects on the participants'

---

* A subtler sign of positive valence and arousal is pupil dilation: the larger the pupils, the happier and more excited the person. During the Renaissance, women would use drops prepared from the deadly nightshade plant to expand their pupils. As a result, when a woman would look at a man, the man would assume that the woman found him appealing. This is why the name for that plant became "belladonna," Italian for "beautiful woman."

conscious feelings: participants who were ostracized felt much lower levels of four fundamental social needs—belonging, control, self-esteem, and meaningful existence. In a follow-up study, participants also exhibited these negative effects when they were ostracized by two computer agents instead of two people—another example of how people respond socially to computers.*

Arousal's link to the body is even stronger than valence's link. When people are very excited, their heart races, their blood pressure increases, and their temperature rises. While we cannot see these responses directly, the body shows their effects. For example, blood is distributed unevenly over the body: when you are angry, your face gets redder as blood rushes to the head, and when you are afraid, your blood rushes away from the head, leaving you "white as a sheet." Because blood flows to the extremities and the muscles of the limbs contract, excited people move more rapidly: they simply cannot stay still, with their body rocking and their fingers and toes tapping as they fidget in their chairs.

Arousing events also kick the cognitive system into active thinking and attention. When you're excited, your eyes constantly scan the environment; the smallest disturbance causes rapid movement in every part of the body. Excited people speak more, more rapidly, louder, and with a relatively high pitch and large range of pitch and volume. When listening to others speak, aroused people interrupt frequently. Very calm people on the other hand are still. They take long glances and move very little and in a measured way. They seem to be immune to small changes in the environment. They speak slowly, softly, and with very little affect.

Surprisingly, just as the brain controls the body to manifest and respond to emotions, the body can convince the brain that an emotion

* Ironically, Williams was unaware of my research when he ran the computer agents condition: he thought that it would show that ostracism is profoundly human and was shocked to find that it was not. It was only while presenting his research that a fellow psychologist pointed him to the finding that computers are social actors. We have since become good friends and colleagues.

is warranted. If the body is experiencing the reactions associated with a particular emotion, the brain recognizes this and reinforces that emotion cognitively. For example, psychologist Robert Soussignan had participants clench a pencil in their teeth in two different positions (forcing them to smile different amounts) while they watched television shows. He found that when participants were forced to smile more broadly, it made the shows seem funnier and better to them. Conversely, walk around with a frown and you will feel sadder. The Seven Dwarfs were right: you should whistle while you work.

Similarly, when the brain determines that you should be aroused, it causes your body to release a number of steroids (among many other responses). And if your body is experiencing a rush of adrenaline, this will make your brain think that everything it encounters is more arousing. Perhaps the most famous example of this phenomenon is a classic field experiment by psychologists Donald Dutton and Arthur Aron. In the study, experimenters watched for men approaching a very rickety bridge.* The idea was that the instability of standing on the bridge would raise levels of adrenaline and arousal. Half of the participants were interviewed by a confederate while they were on the bridge; the other half were interviewed by the same confederate before they stepped onto it. The confederate was an attractive woman who asked the participants to make up stories based on ambiguous pictures that she showed them. She then gave her name and phone number, inviting the participants to call her if they had any questions regarding the project. Consistent with the idea that the arousal caused by the bridge became linked to perceptions of the interviewer, men who the female interviewer approached while on the arousing bridge were much more likely to tell stories with sexual content and more likely to call her afterward.

* This was a field experiment, which has two important characteristics: 1) it is done in a normal setting rather than in a laboratory, and 2) the participants do not know that they are in an experiment! It's often very hard to find the right conditions for a field experiment, but when you can make it work, the results are always compelling.

My son, Matthew, experienced the link between physiology and subsequent emotion firsthand when we went to see the movie *Batman* right after he had used an adrenaline inhaler for his asthma. Even though he normally handles tense movies with aplomb, this time he ran out of the theater and had nightmares for months—his brain mistook medicinal adrenaline for his reaction to the movie! This confusion of bodily feelings with subsequent emotional feelings occurs with calmness as well: after a massage, which triggers the parasympathetic nervous system, even a rocket launch is responded to calmly.

The fact that emotions have a physiological component makes understanding them simpler in another way as well. People typically think of being angry with a colleague versus being angry at a project versus being angry with a table on which you've stubbed your toe as very different situations. Similarly, liking a colleague, liking a memo, and liking free doughnuts seem to be different types of "liking." However, the three types of anger lead to virtually identical reactions from the mind and body; likewise the three types of liking lead to their own virtually identical responses.

In essence, emotions, in terms of the body's reactions, depend on the feeling inside a person, not the external cause of the emotion. Perhaps love, the most other-oriented feeling possible, vividly illustrates this self-centeredness of emotion. If you Google "greatest love songs," you come upon two top-ten lists: one from the popular music network VH1 and the other from the rather conservative New York *Daily News*. If you examine the lyrics of the combined nineteen songs (one song appears on both lists) with a total of more than 750 lines, you see that only 6 lines specifically describe the beloved. One song provides four facts ("long blonde hair," "beautiful lady," "love light in [her] eyes," and "she is wonderful tonight"), and two songs provide one fact ("the sweetest eyes that I've ever seen" and "she is so sincere"). In describing love, you can understand the sentiment despite being told nothing about the object of that love.

In most cases, look at a person and you can unambiguously determine her or his emotional state because all the various physical indicators described above point to the same level of valence and arousal. Sometimes, however, people's emotions are less clear-cut, as they exhibit some signs of happiness and some of sadness or some of excitement and some of calm. This often happens when people consciously try to hide an emotion that they are feeling; that is, their cognitive system tries to take control and to send a different signal from what their body is naturally expressing. For example, angry people's faces are red even if they try to modulate their voice and say, "Everything is fine." Similarly, relaxing one's body to "act casual" can help hide an upcoming surprise party, but the excitement in one's voice may give it away.

How do people interpret contradictory emotional signals? More specifically, does your emotional delivery affect how others understand and believe what you are saying? To address this question, I developed an experiment, along with Ph.D. students Ulla Foehr (now a researcher at the Kaiser Family Foundation) and Scott Brave (coauthor of my second book, *Wired for Speech*) and a group of undergraduates, to study the relationship between what people say and how they say it. Given that I wanted to examine only one channel of communication, I had participants interact with a telephone-based system. This allowed me to zero in on tone of voice without other considerations such as physical appearance or body language.

Experiment: Can You Hear My Emotion Now?

We had participants call a number to listen to two news stories, two movie descriptions, and two health stories. For each category, one of the stories was happy and one was sad. For example, the happy news story described a promising new cure for cancer, especially for children; the sad one described a spate of dead gray whales washing ashore in San Francisco.

The same male synthetic voice presented all the stories. We could manipulate the voice to sound either happy (higher pitch, more pitch range, and rising intonation at the end of phrases) or sad (lower pitch, less pitch range, and falling intonation at the end of phrases). For half of the participants, we made the emotion of the voice match that of the story it was reading: a happy voice read the happy stories, and a sad voice read the sad stories. For the other half of the participants, we mismatched the emotions: a sad voice read the happy stories, and a happy voice read the sad stories. After listening to the stories, participants then completed an online questionnaire that asked how happy or sad the participant thought each story was and how much they liked it.

➤ Results and Implications

People were more optimistic about the cure for cancer when a happy voice described it and were more concerned about the dead gray whales when a sad voice described them. In other words, when the emotion of the content and the voice matched, the emotional message was clear and strong. On the other hand, mismatching the tone of voice muddled the emotion of the content. The results of the questionnaire also indicated that participants preferred the stories with consistent emotional expressions: participants liked the happy stories more when the happy voice read them and the sad stories more when the sad voice read them.

Why would a mismatch between a person's emotional signals (as in the valence of the tone of voice) and what the person is saying (as in the valence of the story being told) affect both the listener's understanding and her or his enjoyment of what is said? Essentially, the mismatch results in significant cognitive work to process the meaning of the message as people's brains try to create a coherent picture from the opposing emotional information (similar to how people with inconsistent personality traits are harder to remember and understand). For example, neuropsychologists Evelyn Firstl, Mike Rinck, and Yves von Carmons monitored people's brains while they listened to short stories, some of which had emotionally inconsistent descriptions of the main character (e.g., both happy and depressed or excited and passive). The emotionally inconsistent descriptions led to activity in the negativity and excitement parts of the brain, probably due to frustration with the lack of success at interpreting the mixed emotional signals.

In sum, people do not combine emotions when they observe opposing signals; instead, they feel confusion and dislike. Failure to recognize this leads to a number of damaging practices. I have seen managers attempt to make criticism seem less negative by giving it with a smile. Instead of softening the blow, this leaves the evaluated person feeling uncomfortable and confused about what the manager wants changed. Similarly, I have seen managers present exciting news, such as a record-setting quarter, in a calm voice so that people will remain focused for the remainder of the meeting. Unfortunately, the mismatch actually creates a distraction: "What is he trying to hide?" "What is the catch?" "Is this actually a bad sign?" While moderating your emotions can be effective and appropriate (which we will discuss with respect to emotion regulation), you cannot balance levels of valence or arousal by mixing and matching signals.

Knowing the signs for positive and negative valence and excited and calm arousal, you can recognize any emotion (as long as the person presents consistent signals). The next question, then, is how these different emotions affect day-to-day life. Is a certain valence and arousal state preferable, and if so, how can you get yourself and others to this ideal state? We address these questions first in terms of valence and then in terms of arousal.

For valence, the social science literature clearly indicates that happiness has a number of positive benefits. For example, the great gestalt psychologist Karl Duncker found that even mildly positive feelings improve thinking and problem solving. When given the task to attach a lighted candle to a wall using thumbtacks so that no wax drops on the floor, happy individuals did much better than sad individuals.* In another study, psychologists Alice Isen and colleagues asked medical students to diagnose patients based on X-rays after first putting them into a positive, negative, or neutral mood. They found that the happy students reached more correct conclusions more quickly than the other groups.

While keeping every single employee happy may seem an unattainable goal, once people are happy, their positive state actually self-perpetuates to a certain degree. This occurs because after an emotion-causing stimulus has come and gone, the physiological correlates of emotion take some time to return to their neutral state. That is, the physical processes of our bodies have much greater inertia than the thinking processes of our brain. Just as forced smiling makes television shows seem funnier, a physiologically happy body leads you to see the world through "rose-colored glasses," which helps sustain your positive feelings. As a result, it takes much less effort to keep generally happy employees happy as compared to trying to cheer up generally morose employees with the occasional extravagant holiday party or social. Similarly, negative events tend to make people find the negative in subsequent events; thus, sadness also self-perpetuates. Consequently, while people usually think of someone who did a great job as a "tough act to follow," it is actually tougher to follow a bad performance: the audience will be looking for the weaknesses rather than for the strengths in what you're doing.

Once some of your employees are happy, this positive valence will actually spread via "emotion contagion," as termed by social psychologists Elaine Hatfield, John Cacioppo, and Richard Rapson. Emotional contagion results from people's tendency to automatically mimic each other's language, facial expressions, postures, and movements. Because of the link between body and emotion, the physical imitation

* The trick is to empty the box containing the thumbtacks, pin it to the wall, and then put the candle in the box.

leads people to feel others' emotions as well. Thus, when you see someone smiling or demonstrating other signs of happiness, you unconsciously imitate them and subsequently feel happier yourself.

Valence is so contagious that you don't even have to encounter it directly for it to affect you, according to research done by political scientist James Fowler and internist and social scientist Nicholas Christakis. They studied a social network of people from Framingham, Massachusetts—connected via friendship, family, spousal, neighbor, and coworker ties—and found that one person's happiness can actually increase other people's happiness out to three degrees of separation. In other words, your happiness is not only related to the happiness of your own friends but that of your friends' friends' friends. Fowler and Christakis even quantified the effect: a happy friend increases your probability of being happy by about 9 percent, and each additional happy friend adds to this effect. In other words, valence has a collective aspect.

Fowler and Christakis also studied online social networks, specifically Facebook photos. They examined the photos of students and of friends in their network. Similar to the clustering effect with happiness in face-to-face social networks, if you are smiling in your profile photo, you are more likely to be central to your network and amid a cluster of smiling friends. Furthermore, statistical analysis of the network showed that people who smile tend to have more friends. "Smile and the world smiles with you" is valuable advice.

Unfortunately, the other half of the saying—"when you frown, you frown alone"—is not true: when you frown, others will unconsciously imitate your frown and become unhappier as well. As discussed in the Praise and Criticism chapter, this effect is exacerbated by the fact that people are not Pollyannas: all other things being equal, people look for the negative rather than for the positive. (This may be the origin of one of the Four Noble Truths of Buddhism: "To live is to suffer.") Thus, stories of "my great boss" lack the richness of detail and the sticking power of "my worst boss" horror stories.

Negative experiences are much better remembered than positive ones, and as noted in the Praise and Criticism chapter, *proactive enhancement* makes events after the negative event more memorable. On the other hand, there is *retroactive interference:* people tend to forget the events that happen before a negative experience. This explains why people frequently have trouble remembering what led up to a negative incident but can readily recall the aftermath. This also explains why learning from your mistakes can be particularly difficult: it is hard to recall what created the problem!

It would seem that all of this provides an easy remedy to people's sadness: just surround them with happy people and happy events, and they will feel happy, reflecting the contagiousness of valence. On the other hand, while anecdotes are the bane of science, I was haunted by a memory of a colleague bouncing up to me and saying, "Let's turn that frown upside down!" Instead of making me feel better, I found myself more irritated and distracted after our encounter. This led me to the question: is it always better to work with a happy person, or when you are sad, is it more comfortable and effective working with another sad person?

Experiment: ---------------------------

Does Misery Love Miserable Company?

To investigate this question, I designed an experiment with colleagues from Stanford (Ph.D. students Helen Harris, Scott Brave, and Leila Takayama) and Toyota (Ing-Marie Jonsson, Ben Reaves, and Jack Endo) that paired participants with a virtual interaction partner (whose emotions we could control) while they completed a task. To ensure that participants felt either happy or sad, we had half of them watch seven minutes of happy video clips—scenes at the beach—and had the other half watch sad video clips—the scene in *The Champ* in which Ricky Shroder cries over his father's death and the scene in *Bambi* in which his mother dies. (If we did not directly induce emotion, we might be studying generally happy, generally sad, or neutral people instead of people who felt happy or sad at the moment.) The participants did not know that watching these scenes was part of the study, as we told them that it was for a separate study involving film clips.

We then had participants use a driving simulator to drive a car down three simulated courses, controlling the car with a gas pedal, brake pedal, and force-feedback steering wheel. Also along for the ride was a "virtual passenger," a recorded voice played by the car. The voice belonged to a female actress and made light conversation with the participant throughout the drive. The passenger's remarks encouraged the driver to talk back. For example: "How do you think that the car is performing?" "Do you generally like to drive at, below, or above the speed limit?" and "Don't you think these lanes are a little too narrow?"

While the passenger said the same thirty-six remarks to all the participants, her tone of voice varied. For half of the happy participants and half of the sad participants, the voice was clearly happy and upbeat; for the other half of the participants, the voice was clearly morose and downbeat. In other words, the four conditions consisted of happy drivers with a happy passenger, sad drivers with a happy passenger, happy drivers with a sad passenger, and sad drivers with a sad passenger.

During the drive, the simulator automatically recorded the number of accidents that each participant had. We also determined how much attention the participants paid to the drive by measuring their reaction time: we told participants to honk their car horn as quickly as possible in response to randomly occurring honks that they heard throughout the course. Finally, we measured people's social engagement with the virtual passenger by recording how much the participant spoke with the agent. After the driving was over, we asked participants a number of questions about their feelings about the car and their driving experience via an online questionnaire.

➤ Results and Implications

Consistent with the benefits of happiness, happy drivers had fewer accidents and paid more attention to the road (reacting more rapidly to the horn honks from the simulator). What about the sad drivers with the happy passenger? Did the happy passenger "cheer them up," thus improving their driving? The simulator results suggest an emphatic *no*. The happy voice in fact worsened sad participants' driving: sad drivers hearing the happy voice had approximately twice as many accidents on average as the sad drivers hearing the sad voice. Sad drivers with the happy voice were also less attentive to the road than those with the sad voice (taking longer to honk in response to the random honks). Attempts to make the sad drivers happy using emotional contagion clearly backfired.

The questionnaire results also suggest that sad drivers were better off with a sad virtual passenger rather than with a happy one. Specifically, sad drivers enjoyed driving more, liked the voice more, and thought that the car was of a higher quality when the virtual passenger was sad. In addition, even though you might think that a sad passenger and a sad driver would avoid conversation with each other, sad drivers spoke much more with the sad passenger than they did with the happy passenger. Conversely, happy drivers enjoyed driving more, liked the voice more, thought the car was better, and spoke more with the happy voice than with the sad voice. In conclusion, regardless of valence, drivers seemed to not only perform better but to feel better with a virtual passenger whose emotion was like their own. (Although we did not have a condition in which the car did not have a voice, other studies suggest that the responses to a silent passenger would be midway between the matching and mismatching voices as far as performance and feelings.) These results require a rewriting of the conventional wisdom. Telling people to "look at the bright side of life" can be off-putting if they are feeling very sad, and "misery loves company" should be revised to read, as the social psychologist Stanley Schachter says, "misery loves *miserable* company."

Why didn't the sad drivers benefit from the happy voice? Trying to process and pay attention to emotions that differ from your own takes a great deal of cognitive effort (for example, you must heavily use both hemispheres of the brain). As a result, you will be distracted, uncomfortable, and perform worse. Furthermore, when you encounter someone whose emotion conflicts with your own, if the person continues to cling to her or his initial emotion, the lack of empathy is hurtful. People seem callous when they continue to be bubbly and upbeat despite your clear signs of sadness or distress.

On some level, then, sad people want you to recognize their sadness. Does that mean that directly saying to sad people, "You seem sad," will make them feel better as well as feel more positively about you? And how bad is it if you incorrectly read someone's emotion and attribute the wrong valence?

Experiment:

Call Me Anything but Sad

My experiment, performed with Ph.D. student Shailendra Rao and researchers from Dai Nippon Printing, focused on how participants would react to a computer telling them that they were happy or sad either accurately or inaccurately. As in the previous experiment, we began the study by making half of the participants feel happy and half of the participants feel sad by showing them movie clips. Then we told participants that they would be working with a Web-based movie and book recommendation system similar to Amazon's, called Monkey Media. We chose a recommendation system as the context because it provided a justification for determining and discussing their emotions.

The participants were told that the system would make its recommendations based on their answer to an open-ended question. For happy participants, the prompt read, "Describe your perfect day," and for the sad participants, "Describe the last time that you were lonely." We chose these particular prompts to amplify the strength of their feelings: thinking about happy events makes people happier, and thinking about sad events makes people sadder. One participant answered the "perfect day" prompt with this story:

> My day would start with a refreshing swim in the pool. I would spend the morning doing something exhilarating with my friends, such as taking an adventurous road trip or going on roller coasters at an amusement park. The afternoon would be more relaxing—perhaps a concert of my favorite singers. My family and I would have dinner together and afterwards I would have a huge bonfire outside featuring karaoke. The day would end with my friends and me sleeping out under the stars.

A sad participant answered the loneliness question with this story:

> I had come to the U.S. around six years ago from my home country. I was a bit sad at the start since I was away from my family. But I didn't feel very lonely at that time, since I had friends at the office. A few months after that, I went to some relatives in Michigan. There I met so many people from my country and quite a few of my relatives. It was like a small home away from home. When I came back from Michigan, I really felt lonely. It hit me for the first time how much I miss being away from home.

After participants inputted their responses, the system pretended to process them for one minute. When the system completed its "processing," it told half of the happy participants and half of the sad ones:

> **Based on your response, Monkey Media has determined that you are happy. Monkey Media has some recommendations for you based on the fact that you are happy.**

It told the other half of the happy and sad participants:

Based on your response, Monkey Media has determined that you are sad. Monkey Media has some recommendations for you based on the fact that you are sad.

In short, the Web site correctly assessed half of the participants' emotional state, while the other half were "accused" of feeling an emotion they didn't feel.

After the evaluation, the site presented random descriptions of ten movies and ten books reflecting various levels of happiness and sadness. We then gave participants a Web-based questionnaire (on a different computer) about their experience with the system, including how intelligent it seemed, how well it performed, how they liked using it, and whether the experience affected their mood.

➤ Results and Implications

For happy participants, the system that accurately described them as happy was clearly better; it (obviously) seemed more intelligent, gave better recommendations (even though they were actually random), was less frustrating to use, and engaged them more. What about the sad participants? Did they prefer the system that correctly described them as sad or the one that incorrectly described them as happy? When it came to judging intelligence, sad participants recognized that accuracy was important: the system that called them sad was smarter than the system that called them happy. However, in all other domains, sad participants preferred the system that described them as happy, even though the assessment was false. All participants, including the sad ones, felt more frustrated using the system that called them sad and felt happier when the system described them as happy.

When in doubt about how to describe someone's valence, err on the side of happy. Even if the person is sad, the pronouncement is effective, albeit inaccurate. This mirrors the flattery research described in the Introduction: people will like you more if you are inaccurate for the sake of positivity. Never tell people they seem sad: they will not feel any better and will feel more negative toward you.

Controlled valence contagion is one way to empathetically cheer people up without explicitly calling them sad. Adjust your valence to be slightly more positive than how sad the other person is feeling, but not so positive that you seem oblivious to her or his sadness. Having established greater similarity, slowly become more and more positive. By making your changes small and incremental, your sad partner will not feel that you are unsympathetic. The increasing happiness will benefit both of you, and as long as you keep a close match, you will be able to concentrate and understand each other better, and will consequently realize greater success.

I learned this lesson—react but don't label—through yet another tragic dating experience. I had noticed that my most suave friend was a master of observation, picking up on even the smallest piece of information that the girl he was dating provided. Not one of her preferences or interests evaded his notice, and he would cleverly incorporate this information as he wooed her. At restaurants, he could suggest exactly what she would like from the menu. If she made an offhand comment about liking zinnias, he would later present her with a bouquet of them. Her humming to a song on the radio would be followed up with concert tickets for the same band.

I wanted the girl I was dating to know that I was going to be as suave as my friend was, so I said to her:

> I really want you to like me, so I'm going to carefully watch everything you do. Whenever you say something, I am going to think incredibly hard about it. I will talk with all of your friends and acquaintances to find out as much as I can about you. I'm going to even look for clues to your subconscious by analyzing your doodles and examining every unconscious movement and word choice. This should guarantee that we'll have a fabulous relationship.

Sadly, she stopped returning my calls.

Besides carefully applying valence contagion to cheer people up, you can also try using humor. While it's obvious that comedy gets people feeling better, being funny typically is not considered a relevant quality in the workplace. Joking around can be seen as a distraction, a waste of time, or a sign that you don't take your job seriously. On the other hand, it is common to see cartoons and comics on office walls and for people to joke with their colleagues. And corporations such as IBM, Eastman Kodak, and AT&T have hired humor consultants to help stimulate creativity, motivate employees, improve teamwork, and relieve stress. What then should we conclude about humor in the workplace?

I set out to design an experiment in which a computer would use humor while working with a person on a task. I felt a rush of doubt: while I had shown that computers could successfully exhibit emotions, "computer humor" seems even more of an oxymoron than "computer emotion." However, discovering that a computer can make people smile would clearly be valuable: if a computer can successfully tell a joke, even the most clueless among us can benefit from the same strategy.

Experiment:

Is Laughter the Best Medicine?

To see how working with someone who tells jokes affects people, Ph.D. students John Morkes and Hadyn Kernal (both are now interface consultants) and I had participants complete a variant of the Desert Survival Situation, the same classic cooperative task used in previous chapters. For half of the participants, we embedded a joke into some of the suggestions; the other half of the participants did not receive any jokes. The humorous comments focused on the items being discussed, but the content of the jokes did not provide information about the relative importance of the items. In other words, the humor was task-related but not task-relevant, providing no additional insight into the task.

For example, in the no-humor condition, the computer only referred to the relevance of vodka as a survival tool: "Alcohol causes dehydration, so any vodka you consume could lead to trouble. You should rate it lower." In the humor condition, the computer's comment included the same information plus a joke: "As everybody knows, vodka is *the* essential ingredient in desert rat flambé! But seriously, alcohol causes dehydration, so any vodka you consume could lead to trouble. You should rate it lower."*

➤ Results and Implications

The results showed that joking with the participant during the task had a number of positive effects. First, the humorous computer made participants feel more positive: humor led participants to don "rose-colored glasses" and they liked the computer and the interaction more. The jokes, lame though they were, even made participants smile more

* The other jokes were:

RAINCOAT: It hardly ever rains in the desert, so wearing a plastic raincoat would just cause you to perspire and dehydrate. (Although, if you filled it with sand, it would make a groovy beanbag chair—complete with armrests!)

MIRROR: The mirror is probably too small to be used as a signaling device to alert rescue teams to your location. (On the other hand, it offers endless opportunity for self-reflection.)

SALT TABLETS: As the *Edible Animals of the Desert* book [another item in the list] says, scorpions and iguanas may need seasoning. Seriously, though, the salt tablets should be ranked lower. Taking salt tablets is like drinking saltwater. It will increase your dehydration.

AIR MAP: Another thing about the salt tablets. A thousand of them are just enough to spell out I'M DYING FOR A SLURPEE in large block letters . . . Finally, the air map: determining your location in a desert will be nearly impossible, with or without the map.

(as we observed by watching recordings of participants' reactions during the study).

Humor facilitated the relationship with the computer as well: humorous-computer participants cooperated more with the computer, as indicated by how closely their final rankings aligned with the computer's suggestions. Humorous-computer participants also responded more sociably, making more friendly and polite comments toward the computer. For example, one participant ended her interaction with the humorous computer by saying, "Nice chatting with you." Thus, using humor while collaborating "greases the skids" of working together.

It's important to note that humor did not distract participants from the task at hand. Humorous-computer participants did not take any longer than the control group to complete the task, which indicates that the jokes did not sidetrack them. Humorous-computer participants also extended similar amounts of effort in terms of the number of original arguments they entered while making their comments back to the computer.

A survey of social psychology research confirms that nonoffensive humor is not only acceptable in the workplace: the right kind of humor contributes to success. Rather than being a significant distraction or a time waster, humor, as shown in a review by professor of education R. Wilburn Clouse and business consultant Karen Spurgeon, can facilitate cooperation and affinity among a group. This shared amusement (which fosters identification) improves performance because of the benefits of teamwork, as described in the last chapter. Furthermore, researchers have found that humor alleviates stress, bonds employees together, and boosts morale and creativity.

While jokes are a powerful way to encourage positive valence, dispense humor with care, as only certain types of humor garner these benefits. For example, people can see disparaging humor as hostile; vulgar or self-deprecating humor can suggest negative attributes about the speaker; and ethnic, racial, and sexual humor can offend as well as become the subject of lawsuits. Avoid intellectual and wordplay humor because, while less risky, it can leave your audience feeling left out if they don't get the joke. The one reliable type of humor is silly and not provocative.*

While you should make sure that you tell the right type of joke, don't worry too much about whether your joke actually gets people to laugh. A follow-up study that we did with a computer that used jokes that weren't funny showed that even if a joke falls flat with the audience, it's just as effective at boosting the performance of the listener—and it doesn't hurt the listener's impression of the teller, either. The benefits of success far outweigh the risk of failure with innocent jokes.

This study also confirms the contagiousness of positive valence. Not only did the computer's humor make people happy, it also encouraged the participants to respond in kind. That is, in their comments, humor participants made jokes back to the computer, even though they knew that it would not understand the humor. For example, in response to the computer's joke about vodka, one participant joked back: "I figured that the vodka would be useful as an antiseptic in case you got injured. Or else you could drink until you went blind."

Where valence is simple and obvious, arousal is subtle and complex: there is no analogue to the rule "very happy is good, very sad is bad" because, when it comes to arousal, very calm and very excited are often worse than their less extreme counterparts. For example, collaborating with highly excited people can lead to frustration and exhaustion. As demonstrated by psychologist Giora Keinan, excited people tend to

* Here are a couple of innocent jokes that proved effective in a different experiment involving computer humor:

- Did you hear about the restaurant on the moon? Great food, no atmosphere.
- How many software engineers does it take to screw in a lightbulb? None. It's a hardware problem.

act without considering all the options and all the consequences of those options. Thus, they want to move on to the next topic without carefully processing or pondering the issues at hand, which can lead to unwise and risky decision making. For example, in a study by psychologists Dan Ariely and George Loewenstein, sexually aroused college students reported greater willingness to engage in unsafe sex as well as in morally questionable behavior in order to obtain sexual gratification.

Most people, with the possible exception of teenagers, seek and feel comfortable with extreme arousal only in small doses: a roller coaster ride can be fun for a few minutes, but after enough runs, it becomes fatiguing and unpleasant. As communication scholar Dolf Zillmann puts it, people like to "manage their arousal," making sure that it does not stay too high or too low for an extended period of time. When people are highly aroused, life is like a music video: random and frantic scenes changing so rapidly that the body tenses and the mind races without having anything specific to focus on. Feeling extremely calm can also be uncomfortable: while lolling on the beach on a warm summer's day feels wonderful, at some point most people become restless and rebel against the abject inertia they are experiencing. And while remaining calm under pressure is admirable, talking with someone who is chronically placid is not very satisfying; it can feel as if the person is not actively listening to, internalizing, or grasping your message. Picture having Wally from *Dilbert* as your boss.

Finding the sweet spot for arousal depends on the particular situation. For example, when you want someone to get on board with your proposal, catch the person in a slightly excited state. Excited people are focused and oriented to action rather than to contemplation: for them, a bad decision is better than no decision at all. As long as there is some logic to your proposal, people will agree to it. Similarly, excited people prefer more active recommendations and riskier alternatives. Therefore, if your customers are excited, have the papers for them to sign readily available and don't force them to read. Make your explanations short: excited people want answers and they want them now! The easiest customers to sell to are happy and excited: they see the bright side of the product and want to actively make those good outcomes continue. Selling to calm people, on the other hand, requires you to challenge your customers a bit to motivate them; otherwise, they will never get aroused enough to say yes.

A highly attentive and action-oriented state can also have its pitfalls. Excited people tend to see the trees and lose the forest; they remember the details of an experience but don't have a general picture of it and don't store concepts and principles in memory. That is why little obstacles will irritate an excited person, while a calmer person will have perspective on the situation and "not sweat the small stuff."

Another pitfall of excitement (and, to a lesser extent, calm) is that it lingers long after its cause. This is explained by Zillmann in terms of "excitation transfer": after an arousal-causing stimulus has come and gone, an activated sympathetic nervous system takes some time to return to its deactivated state. As a result, excitement and calm influence reactions to subsequent experiences, sometimes even after the original event and the valence associated with it have been forgotten. In other words, physiological processes (as in experiencing excitement) are much slower than brain processes (as in consciously thinking about the event and feeling valence). For example, I once saw a feud within a group ensue over the borrowing of someone's stapler while a neighboring group was having a shouting match over a dirty cup left in the sink. When I tried to defuse the situations by pointing out that everyone's behavior was clearly out of proportion, the agitated parties insisted that this was the "last straw" and "more important than you understand." When I investigated further, the only unusual occurrence I could find was the great news that the company had received a long-awaited contract!

How did such a major positive event followed by tiny problems explode into frustration and anger? Although the positive valence of the contract faded, the high arousal lingered and became attached to the stapler and coffee cup incidents, leading to fighting (excited and negative behavior). Another instance of valence changing while arousal remains high is when people laugh uncontrollably at funerals. In their extremely aroused state (initially connected to negativity), even the most mildly amusing event produces a strong reaction. This is immortalized in the episode of *The Mary Tyler Moore Show* in which Chuckles the Clown, while dressed up as a peanut, is killed by an elephant that attacks him. The incident leads to a lot of joking among the newsroom staff ("You know how hard it is to stop after just one peanut!"). Mary is not amused and chides the staff for their lack of respect. However, at the funeral, she inexplicably finds the eulogy hilarious and bursts out into laughter (an excited, positive response). The minister tells the mortified Mary that this laughter is all right and in fact appropriate for Chuckles, telling her to "laugh for Chuckles." At that point Mary starts to sob uncontrollably, her arousal level still high but her valence negative once again.

Avoid presenting aroused people with even minimal negative experiences, as their arousal can quickly turn to anger and acting out, even if they originally were feeling positive valence. Rioting demonstrates this principle. One would think that the destruction of property and the violence that occur during rioting results from extreme negative emotions. However, when interviewed by computational modeler Nanda Wijermans, a Ph.D. student at the University of Groningen, and her advisors, rioters landed along the entire valence spectrum, from very angry (the Rodney King riots) to very happy (riots after a sports championship); what they all had in common was very high arousal.*

The long-lasting nature of arousal can be compounded by the fact that excitement is highly contagious. When two excited people work

---

* An excellent example of leveraging the independence of arousal and valence comes from Marc Antony in Shakespeare's *Julius Caesar.* After Brutus kills Caesar, Marc Antony gives a speech that starts, "Friends, Romans, countrymen, lend me your ears; I come to bury Caesar, not to praise him." As he continues his speech, he gets the crowd feeling extremely happy and excited about how wonderful Brutus is. Antony then begins to insert more and more negative comments about Brutus while still using intense language to keep the crowd's arousal high. By the end of the speech the crowd's arousal is very high while the valence has shifted from positive to negative: a joyous assemblage has become an angry mob.

together, the matching high levels of arousal can create an echo chamber, increasing the excitement of both. You shout and jump up and down more when watching a football game in a crowd at a stadium than when watching it on television, and more when you are watching with friends than when watching alone. Thus, although two excited people can initially work well together, as they are both action oriented, trouble can arise if they drive each other to overly high levels of excitement.

Although excited people are useful for doing things and doing them rapidly, when a careful decision needs to be made, you want calm people. Calm people (as long as they are not *too* calm) carefully process information and separate the wheat from the chaff before making a decision; they do not act prematurely. Calmness helps prevent against groupthink as well as risky decisions, which are particularly anathema to calm people. Furthermore, once a decision is made, calm people will stick with the decision; excited people may change their minds just because they are itching to do something.

Calmness has few of the pitfalls and risks that excitement does because although calm has some contagion effects, the presence of other people tends to increase arousal, an effect known as "social facilitation." In what many people consider the first true social psychology experiment, performed in 1898, Norman Triplett showed that bicyclists rode faster when there were other bicyclists around than when they were riding solo. Later, Robert Zajonc, one of the most important psychologists of the twentieth century, demonstrated the link between presence of others and arousal across a number of animals, including cockroaches, horses, and humans. Thus, you can create a calm environment in yoga classes, meditation groups, or even in a low-key office without worrying about people being driven into catatonia.

How to Quiet the Overly Excited and Stir the Overly Calm

Given the desirability of slight excitement or calm (and the disadvantages of extreme excitement or extreme calm), how can you get people to just the right level of arousal? Never try to argue people out of their feelings by explaining why they should not feel that way: excited people don't respond well to reason, which focuses on thinking rather than on action. Logic-based approaches to handling excitement lead people to feel negative—people dislike being told that they shouldn't feel what they are already feeling—and will increase their arousal while giving them a new target for their excitement—you. These pitfalls are well described by John Gray in *Men Are from Mars, Women Are from Venus,* as men often use this strategy when dealing with upset women.

Approach reducing someone's arousal similarly to cheering up sad people: adopt a level of arousal that is slightly lower than the other person's but not so low that you seem insensitive. Then slowly lower your own arousal level; the other person's level will adjust through contagion. This can be difficult because the other person's higher arousal is more contagious than your calm, so carefully modulate your own arousal. Making a calm person more excited should follow the same strategy and is easier to achieve because, in general, calm and positive emotions are more susceptible to change than excited and negative emotions. That is, the sympathetic nervous system and the right side of the brain are more powerful than the parasympathetic nervous system and left side of the brain, respectively.

For this reason, extreme negativity coupled with extreme excitement is the hardest emotion to relieve. Frustration is one such example, a particularly destructive emotion in the workplace. Once people are frustrated, even very minor negative or arousing events that would normally not be problematic can make them even more frustrated, similar to how arousal contagion can escalate excitement in a group. The following experiment demonstrates how you can manage others' frustration.

Experiment:

There Is Nothing More Frustrating Than Talking About Frustration

Jonathan Klein, a master's student of Professor Rosalind Picard of MIT's Media Lab, had people interact with a computer to test different tactics for relieving frustration. Given a computer's propensity for causing frustration, it seemed an ideal confederate to explore these questions.*

Typically, engineers try to design against user frustration by either troubleshooting a problem after it occurs or preventing problems before they happen. People adopt similar approaches when they encounter someone who is frustrated: they ask what they can do to help solve the person's problem, and if they are responsible, they try to avoid the frustrating behavior in the future. Regardless of whether you are dealing with a person or a computer, these strategies are far from straightforward: the first requires recognizing the problem as well as knowing how to solve it, and the second requires predicting future problems. Instead of tactics that target the cause of frustration, Klein focused on strategies that would address people's emotional reaction, trying to reduce their levels of negativity and excitement.

In the experiment, Klein told participants that they would be testing a new Web-based video game. The game involved a character collecting treasures while navigating a maze. Participants had a chance of winning a hundred dollars if they obtained the best time on the task, incentivizing them to do well. When participants actually played the game, their character would occasionally freeze on-screen while the game timer continued to advance. Unbeknownst to the participants,

* Normally, I would not present a study performed by another laboratory in this format. However, the study looks remarkably like studies performed in my lab (indeed, I wish that I had done the study!), and my former Ph.D. student Youngme Moon and I were heavily involved in the design of the experiment, so I felt that it would be appropriate to make an exception.

the researchers had built in these delays to cause frustration (a likely reaction given the money at stake). After the game ended, the computer asked participants to evaluate the game by answering a series of questions.

To explore tactics for alleviating negative/excited emotions, Klein had the computer present participants with one of three different online questionnaires. For one-third of the participants, the questionnaire did not allow participants to report problems they encountered or to describe how they felt about their experience. They could only respond to multiple-choice questions with no opportunity to describe their feelings. A second group of participants were invited to "vent" through answering open-ended questions that asked them to write about the frustration they felt and the problems that they had encountered. For example, in the vent condition, the questions included:

> If there were any delays, do you think they affected your game?
>
> How frustrated do you think you got playing the game, all things considered?

The third strategy for relieving frustration was emotional support. Specifically, the computer asked participants the same questions as the second group of participants, but after the computer asked each question, it gave text-based feedback based on the user's reported frustration level, playing the role of an "active listener." For example, the computer acknowledged that it had "heard" the participants' frustration—"Wow, it sounds like you felt really frustrated playing this game." Implying that it wanted to "understand" their frustration, the computer also gave participants opportunities to correct what it had "heard"—"Is my judgment of your feelings about right?" It also sympathized—"That must feel lousy. It is no fun trying to play a simple game, only to have the whole experience derailed by something out of your control"—and even took some responsibility—"This computer apologizes to you for its part in giving you a crummy experience."

To test which of these strategies (nothing versus venting versus emotional support) most reduced people's negative and excited feelings, the researchers then asked the participants to play the same game a second time, but this time they removed the delays so as to avoid additional frustration. In this round, the participants could quit anytime they wanted simply by pressing a large QUIT button. The idea was that if people still felt frustrated from playing the game the first time—if the approach to questioning did not alleviate their frustration—they would quickly decide to quit. This allowed the researchers to quantify the effectiveness of the tactics.

➤ Results and Implications

Participants who worked with the "emotional support" computer played the second game much longer than did participants in the other two conditions. Actively acknowledging and addressing people's emotional states alleviated the high negativity and high excitement associated with frustration. In other words, people feel better when you show that you have heard them, understand their feelings, and sympathize.

While being a good listener is commonly regarded as a good way to deal with frustrated people, this experiment shows that just listening doesn't actually help. Allowing venting (without any support) was no different than not allowing venting at all. Both groups played the second game for only a very short time, showing they still felt negative toward it. As Klein and other researchers have concluded, venting alone probably helps people recall the situation and how frustrated they became with it. This causes them to dwell on their negative and excited feelings (similar to how recounting a happy or sad event will make one feel happier or sadder). Without coupling venting with support that acknowledges the person's emotion, those feelings remain negative and unresolved. Thus, inviting people to simply vent does little or nothing to help them get past their feelings.

The present study also provides more guidance on how and when emotion should be discussed. While in the previous study participants heartily disliked the recommendation system that called them sad (even when the description was accurate), in this study, recognizing users' frustration had positive benefits. Thus, you should describe someone as frustrated or angry (negative and highly aroused) only if you are also supplying support to help alleviate that emotion. The rules for relieving negative valence with low arousal, i.e., melancholy (or, in its most extreme form, depression) are similar to those for frustration or anger. That is, you should provide emotional support rather than simply encouraging venting: active listening, empathy, and sympathy are the best bet.

There is a subtle but important difference between frustrated and melancholy people. When people are frustrated, their high arousal makes them want to do something about their negative emotion. As a result, they often focus on the supposed source of their negative feelings. Conversely, melancholy people are not interested in acting against the source of their problems (or anything else). Does this difference in orientation affect how you should console negative-excited people (those who are frustrated or angry) versus melancholy people?

Experiment:

Too Bad If You're Sad; Assign Blame If You're Mad

To explore this question, my Ph.D. student Yeon Joo and I had people use a car simulator. Once again, the driving task allowed for measurable performance results and a natural situation to converse with a passenger (in this case, one who would be trying to make participants feel better). The simulator had a 140-degree video screen that provided the same view a driver would have looking through the windshield. The participants sat in the front half of a Ford Thunderbird (we had to cut off the back half of the car because it wouldn't fit in our laboratory). Sitting on the passenger side of the dashboard, the simulator had a "passenger" for participants to interact with: a six-inch robotic head that would turn toward the driver whenever it spoke and otherwise would turn back to look at the road. We used the robot as the tour guide so that the comments would be independent of the car (much as a computer asking opinions about a different computer creates the opportunity for greater objectivity than does a computer asking about itself).

Before using the simulator, we made participants feel either very angry or very melancholy via video clips, similarly to how we made participants feel happy or sad in other experiments in this chapter. To strengthen their feelings, we gave the angry participants unsolvable word problems to solve, and we gave the melancholy participants a very sad story to read about an infant's incurable disease. Participants were then told that they would be taking a fifteen-minute guided tour of Smithtown, "a charming city with a population of 471,000 people." During the drive, they would be introduced to the historic government building, the water tower, a few restaurants, and numerous other sights.

For the study, we introduced obstacles for the participant (creating negativity) in order to give the tour guide an opportunity to empathize with the participant. Drivers encountered four hazards: a construction zone that was too small for the car to fit through, a curve that was so slippery it forced the car off the road, a group of children that suddenly crossed the street, and a police car that showed up in the rearview mirror only to crash into the driver. After each of the hazards, the "tour guide" spoke to the driver about it, displaying empathy that reflected either an action-oriented or a passive approach to half of the angry participants and half of the melancholy ones. Action-oriented empathy blamed a source of the problem that could be changed or otherwise dealt with. In this case, the car identified a particular person as the cause of the hazard: "The construction worker should not have put those barrels so close together. He should have laid out the construction zone more carefully. He should know that he made it impossible for drivers." The more passive version focused on the general situation and blamed the "nature of things," implying that the hazard was out of anyone's control and unavoidable. In this case, the car talked about the situation in general: "Construction zones are very difficult to navigate. They are often difficult to avoid. Construction zones make it very challenging for drivers."

➤ Results and Implications

The results confirm that angry and melancholy people benefit from different types of emotional support. For angry participants, the tour guide that identified a person as a cause of the hazard seemed friendlier and more supportive, even though one might think of blame in general as harsh and uncaring. Angry people also found the tour guide that blamed a person more accurate, dependable, and trustworthy. On the other hand, melancholy participants preferred the tour guide that acknowledged the negative occurrences but implied they were unavoidable.

The style of emotional support even affected participants' driving, a sign of how effective each strategy was at making participants less upset (and consequently safer drivers). The person-blaming tour guide helped the angry participants exhibit fewer poor driving behaviors such as speeding and lane crossings, while the situation-blaming tour guide facilitated safe driving for the melancholy participants. As with miserable people's negative response to happy company, when active and passive perspectives are mismatched, dislike, distraction, and poor performance ensue.

This study provides nuance to how one should provide social support. For frustrated people, include explicit recognition of and discussion of actionable sources of their negative and excited feelings. These people are driven by their desire to act; attributing causes to things that cannot be affected can feel dismissive. For melancholy people, imply, subtly, that there is nothing to be done. Calm people don't want a call to action, as it conflicts with their desire to remain passive.

While emotions evolved in many animals to speed responses to the environment, humans are unique in that we can take control of and create our own emotions. We manage (or at least try to manage) our emotions and how people see them all the time. Indeed, people are expected to hold back from giving "too much information" (TMI) and to not let their emotions "run away with them." For example, when your boss unfairly criticizes you, you don't (usually) run away or punch someone in the face. When people feel sad at work, they don't usually cry or slump inertly onto their desk. And when you receive a raise, it is inappropriate to jump up and down and hug everyone (although this response is perfectly acceptable when you win on television game shows).

Regulating your emotions is important because displaying the right emotion at the right time is crucial for getting along with others. It is not surprising, then, that when emotion psychologist James Gross at Stanford interviewed hundreds of people about times that they had recently regulated their emotions, 98 percent of the episodes took place in the presence of others. He also found that people regulate negative emotions the most, specifically controlling sadness, anger, embarrassment, anxiety, and fear. In fact, the least hidden negative emotion (disgust) was regulated more often than the most regulated positive emotion (pride).

While controlling the appearance of your emotions is very important, it is also very difficult. The extremely tight linkage between the feeling part of emotions and the body means that it is hard to turn off "tells" in your eyes (pupil dilation), eyebrows (raised or lowered), the mouth (smiling or frowning), skin (flushed or pale), extremities (motion in the arms, legs, hands, and feet), and voice (changes in pitch, volume, and speed). The difficulty of stifling these telltale signs doesn't keep most people from trying—*suppression* is a common approach. However, by and large, research has shown that suppression is not that effective. First, it never is perfect: the inconsistencies between the manifestations that are suppressed and the ones that leak out (almost every poker player has a tell) suggest that the person is inauthentic. And because suppression requires managing emotions as they occur, you have to work hard to do it, which draws cognitive resources away from accomplishing other goals. As a result, experiments conducted by Gross and psychologist Jane Richards show that suppression leads to increased work for the heart, worse memory for social information (such as names or facts about individuals) and for conversations, and less satisfying social interactions. Gross notes that people who frequently suppress their feelings have lower levels of satisfaction and well-being, less life satisfaction, and a less optimistic attitude about the future, consistent with their avoidance and lack of close social relationships and support.

In sum, once emotions arise, it is often impossible and inadvisable to aggressively hide them. Gross's research instead suggests that to avoid manifesting a highly emotional reaction to an event, *stop yourself from experiencing the emotion in the first place!* This can be easier said than done, as the primitive parts of our brain are continually trying to assign emotions to events. However, before those automatic reactions engage the higher-level parts of the brain, you can intervene. Using "cognitive reappraisal," you can reframe how you interpret a situation to reduce the event's impact on your emotions.

Psychologists Joseph Speisman, Richard Lazarus, and Elizabeth Alfert launched the field of cognitive reappraisal in a study where they showed a twenty-minute film about a terrifying surgical procedure performed by "primitive peoples" on young boys. Half of the participants were shown the film with a voice-over that described the events with an air of scientific detachment. This framing dramatically decreased the arousal of the experience, as measured by heart rate and amount of sweat on the hands. In another study, merely telling people that the procedure was neither painful nor dangerous (although it was obviously both) and that the boys looked forward to it as a rite of passage was dramatically effective at calming participants.

Cognitive reappraisal is commonly used when sitting in a horror movie. If your reaction to the film becomes overwhelming, you might tell yourself "it's only a movie, it's only a movie" to reduce your level of arousal.* Similarly, when someone gets the last cookie in the lunch line, you might ameliorate your annoyance by thinking about the calories saved or how instead you can have an even better treat at dinner.

Experiment: They're Not Evil, They're Misunderstood

To determine the effectiveness of reframing in a highly stressful environment, I created an experiment with my Ph.D. student Helen Harris and a team of undergraduates using a car simulator. Drivers experience numerous, discrete events that can spur frustration and anger, making driving well suited for exploring whether cognitive reframing helps alleviate extreme negative/excited emotions.

Participants were told that they would be driving through cities and on highways. They were also told that while they should drive safely, they should try to get to their final location as soon as possible. To push the driver to the extremes of emotion, as in the previous experiment, we introduced common, frustrating situations that occur while driving: evading an erratic cyclist, getting cut off by another car, confronting an enormous amount of traffic, and so on. After each incident, the voice in the car adopted one of two approaches. Half of the participants were encouraged to cognitively reappraise the situation. For example, when the driver was cut off by another car, the cognitive reappraisal participants were encouraged to reframe the situation: "The design of that car makes it difficult to see you; otherwise, the driver would not have chosen to change lanes." In other words, the car pointed out that because the events were not directed at the driver and

* My first book, *The Media Equation*, explains why your brain doesn't respond, "You idiot. Of course it's only a movie. What else did you think we were doing sitting in a theater surrounded by a bunch of people?"

were inevitable, they did not warrant frustration (negative/excited). The other half of the participants simply heard an acknowledgment of the situation: "Being cut off is one of the many difficulties in driving."

➤ Results and Implications

The attempt to defuse negative and aroused emotions by encouraging cognitive reappraisal was extremely effective. Participants who heard the messages that reframed the situation felt more relaxed and more positive about driving. More important, those drivers who were encouraged to reappraise their situation drove much more safely than those who experienced their frustration without having it short-circuited.

Recent research by Gross and his colleagues suggests that your attitude toward emotions can encourage or discourage others to use cognitive reappraisal. Bosses can explicitly discuss with their employees how to best manage their emotions and to rationally evaluate their emotional response. Conversely, bosses who view emotions as dangerous and focus on avoiding and minimizing emotions encourage employees to use suppression, with negative consequences.

Gross's research suggests that, through practice, you will get better at cognitive reappraisal. So when you begin to experience the rush of an undesirable emotion, immediately scrutinize the situation to come up with an understanding that lessens the threat to your goals. By unlinking the situation from your goals, you can alleviate negative and arousing emotions. Make sure to reinterpret situations in a realistic way based on facts and evidence. For example, if you miss a deadline for a project, you may start feeling anxious, thinking that, "This throws the whole project off," "We might not be able to recover from this," and "Everyone will blame me for this project failing." In reappraising the situation, try to rationally assess the situation and your initial thoughts about it. For example, think, "We can still make up the time. I can work overtime and get the project back on track" or "How can I enlist others to help me get this done?"

In contrast, telling yourself to feel better, being falsely optimistic, or rationalizing the problem are less effective because, like suppression, they are based on denial. They can result in an invalid reinterpretation of the situation, which will only reduce emotion in a superficial and temporary way. For example, thinking, "Falling behind is not such a big deal. Probably no one will even notice it," may make you feel better for a little while. However, that relief will be short-lived as you probably will quickly encounter facts that prove the rationalization false (for instance, when missing the deadline results in your boss calling a meeting to hold you accountable). Rationalization, because it encourages ignoring the problem, can hinder you from taking action that would help fix the situation and resolve the emotions you are feeling.

Because cognitive reappraisal can happen before (rather than while) someone fully experiences an emotion, it has many advantages over suppression. Gross has found in several studies that everyday use of reappraisal results in feeling more positive and fewer negative emotions overall. People who use cognitive reappraisal are more satisfied with their lives and are more optimistic. Reappraisal also results in better social interaction because if you successfully reappraise your situation and feel better, you can then concentrate on the task at hand rather than on your own unpleasant emotions. Consequently, reappraisers have more successful lives.

➤ Although some people might pride themselves on understanding the nuance of hundreds of emotions, all emotions boil down to happy versus sad (valence) and excited versus calm (arousal).

➤ Each emotion is linked to certain physical reactions: emotions are felt in both the brain and the body. Manipulating your expressions of an emotion can lead you to send mixed signals, which people find confusing, dislikable, and suspicious.

- Happiness is the best policy: happy people actually work better, think better, drive better, and even make other people happier. Telling people jokes can make them feel happier (and more positive about you) without necessarily detracting from their work.

- Happy people like happy people, but misery loves miserable company. Match people's valence when you work with them, and everyone will be more comfortable and perform better.

- Emotions are contagious: surround yourself with people who have the emotions you would like to have. Emotion contagion must be used gradually when you want to change people's emotions.

- People try to hide negativity, so do not accuse someone of being sad, frustrated, or angry. Instead, actively listen and empathize. Don't try to reason them out of their feelings or just listen to them vent.

- Use cognitive reframing rather than suppression to regulate your own emotions. It's better for both your health and your relationships.

CHAPTER 5

Persuasion

The ability to persuade others ranks as one of the most valuable skills to learn, particularly in the workplace. Employees need to convince management that they deserve more money, managers need to persuade workers to find the corporate vision compelling, and salespeople must get customers to buy their product. Indeed, it can be argued that almost every conversation and communication in the workplace involves an attempt to persuade others: to select one option over another, to dedicate themselves to a particular goal, or to pay attention to someone or something.

In my early days of consulting, I thought I didn't need to try to persuade people because "the facts speak for themselves." I would simply gather all of the relevant data, analyze it carefully, and present it clearly and objectively. Who could object? This naïve assumption was quickly corrected by a client's response to the results from one of the largest and most rigorous studies I had ever performed.

The goal of the study was to select the best pictorial agent from a set of possibilities to be the "face" of a software application's support function. My group obtained a large random sample from across the United States, Germany, and Japan and gathered a great deal of quantitative data from each participant. By asking questions beyond the typical "How much do you like this character?" we developed an extraordinarily rich set of models using techniques including cluster analysis, multidimensional scaling, and multiple regression. Each of the models we generated pointed to the same character: a clear and unambiguous choice that had a compelling personality, fit the product, matched all of the branding consideration, and was easy to animate.

It was so rare and so satisfying to have such unambiguous results that my presentation of our final recommendation ended with: "In all my years of consulting, I have never seen a more compelling and obvious choice. I can honestly say that everything points to one and only one conclusion: you have found the perfect character! Congratulations!" This was a true triumph of science.

The group was cheering and smiling as I pronounced this verdict; I left the room sure that I had done my job. A few days later, I got a call from one of the managers of the company. To my utter disbelief, he told me that they were having second thoughts about my recommendation. He had shown the picture of the agent to his wife and something about it had bothered her. He was now convinced that the group had to go back to the drawing board because of her "gut feeling." I told him that I appreciated intuition as much as the next person, but it was simple: there was indisputable evidence. The discussion went on for another hour, with him talking about "instinct" and me trying to persuade him with the memories of the beautiful graphs, charts, and mathematical models. In the end, the company exercised its "judgment" and chose another character that they had a better feeling about. (The fact that their selection was a flop was no consolation.)

The experience taught me that being persuasive requires more than just having truth on your side. The wide variety of sayings about persuasion also reflects this: "Speak softly and carry a big stick"; "If you wish to win a man over to your ideas, first make him your friend"; "He who wants to persuade should put his trust not in the right argument but in the right word"; "The best way to persuade others is with your ears"; "Power is the most persuasive rhetoric"; and so on.

What these sayings have in common is that they tell persuaders how they should act or the role that they should play. In other words, when people decide whether to accept or reject a suggestion or idea, they take into account the source of the information. Regardless of what you are trying to be persuasive about, who you are and your relationship to your audience play a key part in your success. These "beyond the message" effects are grounded in human relationships and thus susceptible to systematic study using computers. Using this approach, I have uncovered the characteristics and strategies that can make you more persuasive, no matter the core content of your message.

While persuasion has been a focus of the field of rhetoric since Aristotle, the modern study of persuasion was launched in 1951 with a paper by one of the founders of the field of communication, psychologist Carl Hovland, and his Yale colleague, psychologist Walter Weiss. Hovland and Weiss argued that all persuasiveness could be understood in terms of two issues: expertise and trustworthiness. Expertise, discussed in the first half of this chapter, describes whether someone is *worth* listening to; that is, how intelligent and knowledgeable the person is about the subject at hand. Trustworthiness, described in the second part of the chapter, addresses the question of whether someone *should* be listened to; that is, whether the person has your best interests at heart.

Everyday experience shows that intelligent or expert people are persuasive. If a person knows what she or he is talking about, especially if that person knows more than you do, you will assume that her or his comments are accurate. Plato articulated this idea long ago when he advised that people should accept ideas from the "wisest." How, then, do people decide who is "wise"? The most obvious (and impractical) way might be to give every person you encounter on-the-spot IQ tests, SAT questions, or Mensa challenges, but certainly many types of intelligence exist that are not easily measurable yet are relevant. Absent definitive and extensive evidence, people use less pointed strategies to assess by whom to be persuaded.

One common indicator that someone should be listened to is that she or he is labeled an "expert." People do many things to certify their intellectual competence: go to college, earn degrees, obtain an impressive-sounding title, associate themselves with established organizations, and/or get recommendations from respected individuals. *The Wizard of Oz* presents a satirical example of how much influence "proof" of intelligence can have. Because the Scarecrow was unhappy that he didn't have a brain, the Wizard gave him "a doctorate of thinkology." The Scarecrow then impressively announced, "The sum of the square roots of any two sides of an isosceles triangle is equal to the square root of the remaining side." This manifestation of intellect—though mathematically incorrect—coupled with his "Th.D." made the Scarecrow the clear choice as the new leader of Oz.

Experiment:

Can Anyone (and Anything) Be a Specialist?

The *Wizard of Oz* example playfully highlights the power of labels and suggests that even those obviously lacking in intelligence benefit from them. This contradicts the general assumption that it is not the markers of intelligence themselves but what they represent that garners respect and makes you persuasive. That is, labels make sense only as a way to summarize your accomplishments—if you received X degree at Y university, you must have learned a lot, and if you are in position Z at a company, you must be competent and effective at what you do.

I was discussing these ideas about labels over coffee with a colleague, Stanford professor Byron Reeves, an expert on the psychology of television content. After mentioning to him the Scarecrow story, Reeves jokingly said, "If you want to see if labels are truly powerful, you need to study something that is just as brainless as the Scarecrow: a television set. Obviously, televisions don't know anything about what they are showing, so if people think that content from an 'expert' television is better than content from a 'nonexpert' television set, that would really be showing something!"

We both initially laughed at this suggestion, but I later realized that while it might not make sense to describe a television as an expert, we could describe it as a "specialist." The term "specialist" refers to those who have a great deal of experience, training, knowledge, and judgment about a particular subject domain. Given specialists' expertise, their comments, as those from the "wisest men," are very compelling and persuasive. For example, if someone has a brain tumor, she or he will likely take the advice of a brain surgeon over that of a general practitioner, whose advice in turn will be taken more seriously than that of a random person on the street. Similarly, the Web site epicurious.com is seen as having better (although probably more challenging) recipes than About.com. Generalists suffer from the "jack-of-all-trades, master of none" perception. Thus, when individuals can claim a specialization—through extra training, for example—it provides a reasonable basis for finding them more persuasive in their domain of expertise.

Consequently, we decided to design an experiment with "specialist" televisions. Specialist television *stations* exist, such as CNN for news, Comedy Central for humor, and Cartoon Network for animation. Could we create specialist televisions that showed only one type of content? After pondering the possible marketing campaigns ("A television as funny as its shows," "What knows more about news than the Sony Informer TV?"), we decided that the idea might be crazy enough to work. We could then determine whether a mere label of "specialist" could make someone or something that obviously could never be a true specialist—even a television set—more persuasive.*

---

* We knew that the probabilities of success were so low that it would be unfair to formally assign someone to this ridiculous project, so we decided that we would foist it upon the next student who walked through the door. As luck would have it, five minutes later, Glenn Leshner, a first-year Ph.D. student (who we assumed had time to waste as compared to an advanced Ph.D. student), walked into the coffee shop. We leapt up and said, "Have we got a fantastic study for you!" (We had made a pact not to tell the student that the study was ludicrous.) Fortunately, the study worked, and Leshner is now a tenured professor at the University of Missouri in communication and journalism.

For the study, we brought people into the lab to watch news and entertainment shows. Each participant watched four "hard" news stories (the stories were about a wounded police officer, a fraudulent business, a recent book about suicide, and the closing of a military medical center) and four segments from network situation comedies.

Half of the participants watched both types of content on a "normal" television. We told these participants that it was an ordinary TV that showed both news and situation comedies; to emphasize this, we put a sign on the TV that read NEWS AND ENTERTAINMENT TV. The other half of the participants watched the shows on our "specialist" TVs. The first television they sat down in front of had a sign that read NEWS TV. We told these participants that the television was perfectly ordinary but that we used it only to show news; we then had them watch the news segments on it. They then moved across the room to another chair that was placed in front of a television with a sign reading ENTERTAINMENT TV; we told participants that this television was only for situation comedies.

After they viewed each program, participants filled out a questionnaire that asked them to rate the overall quality and likeability of the clip. If the clips seemed better when presented on the specialist TVs, the label would have indeed influenced perceptions of content. To determine if the specialist label made the news seem more "newsy," we asked participants to rate how important, informative, interesting, and serious they found each segment. Similarly, to determine whether the situation comedies on the specialist TV were more entertaining, we asked participants to rate how funny and relaxing they found each segment.

➤ Results and Implications

Even though all participants watched the exact same content and everyone knows that an actual television set has no effect on the content it shows, participants liked the segments presented on the specialist TVs more. The obviously irrelevant label also made the content seem better in terms of representing the "essence" of the genre. Specifically, participants rated the news on the News TV more newsworthy in every respect as compared to news on the generalist TV. Similarly, participants rated the situation comedies significantly funnier and more relaxing on the Entertainment TV. The labels were influential even though every participant insisted when asked that it made no difference whether they saw the content on a single generalist TV or on two specialist TVs. Simply by being labeled a specialist, you will be perceived as more compelling, even if the label is obviously gratuitous and irrelevant.

Because the results were so surprising, we wanted to make sure that they weren't a fluke. We conducted a follow-up study where participants watched twelve news stories on a single television. This time, the segments were ostensibly broadcast on different networks, ranging from "news-only" ones such as CNN to "information/entertainment networks" such as ABC and CBS (in fact, all of the segments were taken from local news shows from other parts of the country so the participants would not have seen them before). We again found that the news-only label made a difference: people found the content from a news-only network more important, informative, and enjoyable than content from a generalist one.

Even more striking, participants rated the picture quality higher (clearer and more vibrant colors) for the programs shown on specialist networks, even though all the participants watched the same images on the same screen. Thus, the "specialist" label influenced not only the participants' assessment of the content but even their perception of the visual quality, a completely unrelated dimension. This is an example of a more general phenomenon called the "halo effect," introduced into the psychology literature by social psychologists Richard Nisbett and Timothy Wilson. In their seminal study, participants interacted with a teacher with a European accent who behaved either warmly or coldly. Participants who had the warm instructor rated his appearance, mannerisms, and accents as appealing, whereas those who experienced the instructor behaving coldly felt that the same characteristics were annoying. Although the label of warm or cold affected their judgments of the unrelated characteristics, the participants got the causality backward: they thought that the characteristics made the instructor seem warm or cold. In a similar fashion, once you are labeled an expert, everything you say will seem smarter, even though people may end up attributing your perceived expertise to the "obviously intelligent" things that you said.

There are a number of ways you can leverage the persuasiveness of arbitrary specialist labels. First, use relationships with other people to mutually enforce the label. For example, in a project group, label each member as an expert on a different part of the project, regardless of whether she or he has special expertise or not. Then, all e-mails or memos directed outside the team that pertain to a particular part of the project should come from the specialist in that area, no matter who actually writes it. Similarly, in replies to questions in a given domain, members should refer to the specialist. While the label of specialist within the group can disrupt group unity, exhibiting specializations to those external to the group is always powerful.

You can also set the terms for discussion in order to position yourself as the person who knows things. For example, when you disagree with someone's proposal but do not have a good argument to put forward, you can continually ask questions relevant to your domain of expertise. It is likely that the proposer will be unable to answer your questions, thereby making you seem like an expert.* Psychologist of the practice of science Michael Mahoney makes a compelling argument that this is the primary motivation for the dreaded Ph.D. orals. After the Ph.D. student has finished an enormous amount of coursework, research, and writing, rigorously evaluated and approved by a wide range of scholars, the student must face a group of faculty who

* A tongue-in-cheek recommendation for making an outrageous claim: preface your remark by saying, "It is unbelievable yet true that . . ." You will gain extra credibility because knowing "unbelievable" or "surprising" things makes you seem more expert. Also, knowing something "true" highlights your insights into what the audience might not know.

get to ask anything they feel like. Those who wish to torture the student can ask the most trivial factoids, expressing horror when the student does not know them. This reminds the student that despite her or his accomplishments, the faculty will always be superior.

Why Are Labels So Powerful?

Why are people such suckers for the label of expert? As summarized in a classic article by Harvard social psychologist Daniel Gilbert, a large amount of psychological literature finds that people automatically assume that everything that they hear is true; it takes a great deal of thought and effort to decide that something is false. There is an evolutionary logic to this: if people had to heavily scrutinize every suggestion they encountered and everything they observed, their brains would not be able to handle the burden. Thus, when people are distracted or confused, they are particularly likely to accept a remark or suggestion—their brain is just too busy to evaluate its validity. Skepticism takes significant effort; people are born to believe. This explains why none of our participants in the first television study, despite our probing, considered it absurd or strange to see a television labeled "specialist": they had no motivation to question.

Another reason people put so much stock in labels is that uncertainty makes people very uncomfortable. Accepting someone's label gives people a clear and unambiguous take on whether to accept or reject the person's suggestions. People are so anxious to get rid of ambiguity that even when they hear information whose accuracy cannot be immediately assessed, they look to confirm rather than disprove, as discussed by Gilbert. For example, if someone tells you that Joe is very kind, you will interpret everything Joe does with the assumption that he is trying to help others. Similarly, if someone tells you that Joan is dishonest, you will look for proof that she is dishonest but not for evidence that she is honest. This tendency to look for confirmation is also behind the phenomenon of television viewers confusing actors with their roles. This explains how in the classic commercial, actor Peter Bergman (best known for his role as Dr. Cliff Warner in the daytime soap opera *All My Children*) compellingly spoke to the medicinal benefits of a popular cough syrup even while prefacing his remarks by saying, "I am not a doctor, but I play one on TV."

Stereotypes about the competence of different societal groups can also result in people assuming that individuals are expert or inexpert independent of their actual knowledge and abilities. For example, people often presume that women are knowledgeable about shopping and more skilled at cooking, and that men are knowledgeable about sports and more skilled at carpentry. Being associated with a social identity (gender, ethnic group, nationality, and so on) that stereotypically engages with a certain category of information can make someone seem more expert in it as well. While stereotypes can be destructive, the sad truth is that they persist, as demonstrated by the German drivers described in the Introduction who would not take directions from a GPS device with a female voice.

Experiment:

You Look Like a Person Who Knows About This

When challenged for harboring stereotypes, people frequently point out that although stereotypes should not be applied to particular individuals, it is not unreasonable to draw general conclusions about particular groups. Independent of any genetic claims, people from a given group generally spend a great deal of time interacting with others like themselves. For example, as described by developmental psychologist Eleanor Maccoby in her classic book, *The Two Sexes: Growing Up Apart, Coming Together*, in all cultures, the majority of people spend the majority of their time with people of the same gender, and even very young children show a striking tendency to segregate by sex when choosing with whom to play. Furthermore, all cultures prescribe markers of gender, thereby making one's own gender more similar and the opposite gender more different than gross morphology alone would dictate. Men and women receive different prescriptions on how to dress, which roles in the workplace and the home are more or less appropriate, which leisure-time activities are desirable, and so on.

Gender is also distinguished via language. In general, women in conversation tend to focus on how things relate and on the interaction itself, highlighting interpersonal aspects and personal feelings. As a result, they tend to use more personal pronouns—*I, you, she, her, their, myself, herself*. On the other hand, men tend to focus on objects and details about them—*the, that, these*—and quantifiers—*one, two, some, more*. Women also express more concern for their listeners and tend to compliment and apologize more. Men and women even use the same word differently: women more frequently use "they" to refer to living things, while men more frequently use "they" for inanimate things.

Thus, the argument goes, gender stereotypes are not unreasonable because they capture consistent differences between one gender and the other. To separate this cultural argument from the mindless application of stereotypes, I decided to study people's responses to different genders of computer voice. Because the computer (and the voice) obviously would not have grown up in an environment with others of the same gender, it would be ludicrous to think that the computer's expertise was based on gendered experiences (as opposed to just based on stereotypes). Thus, I could expose the power of stereotypes by seeing how well male versus female voices commenting on stereotypically male or female topics persuaded people.

For the study, performed with Omron researcher Yasunori Morishima (now a professor at International Christian University in Tokyo), my postdoctoral fellow Courtney Bennett (now a researcher at the Packard Foundation), and my Ph.D. student Kwan Min Lee (now a professor at the University of Southern California), we built an auction site similar to eBay. An auction context was ideal because each item comes with a short description that the gendered voice could read aloud. Also, an auction site allowed us to plausibly present a wide range of products that could have both stereotypically female and male versions. Specifically, we created four categories of products, each of which included a stereotypically female item and a stereotypically male one, such as encyclopedias of sewing and firearms, respectively.

Participants clicked an audio link to hear the description of each item read by a "spokesperson." Half of the participants heard all of the descriptions read by a female voice; the other half heard them read by a male voice. To make the absurdity of stereotyping absolutely clear, we used computer-generated voices that varied only in pitch: the voices sounded more like male and female Martians than anything human. After they were presented with each item, participants were asked about their feelings about the product, the pitch, and the spokesperson.

➤ Results and Implications

Did participants evaluate the products differently based on the stereotypical expertise of the speaker? Very clearly. For example, when the gender of the spokesperson matched the stereotypical gender of the products described, participants found the pitches more persuasive, as reflected in the participants' willingness to buy each of the items. That is, the "female" voice did a better job selling the stereotypically female products, while the "male" voice did a better job selling the stereotypically male products. In addition, when voice "gender" matched product "gender," participants reported that the descriptions seemed more accurate. In other words, matching the gender made the descriptions themselves more believable and the voices selling them seem more expert. Given that the voices were not human, the speakers obviously could not know anything about the content nor use the products! Nonetheless, stereotypes played an enormous role in making the spokesperson more persuasive.

The connection between gender and stereotypically male or female knowledge is so strong that the gender of the spokesperson even influenced perception of the "gender" of the product. The female spokesperson made all products seem more feminine, while the male spokesperson made all products seem more masculine. In addition to the spokesperson skewing the perceived gender of the content, the content also affected how people perceived the gender of the voice: spokespersons of both genders sounded more feminine when describing female products and sounded more masculine when describing male products.

This experiment confirms that stereotypes are still rampant and mindlessly applied, even when unambiguously unjustifiable. How then can they be overcome? One common approach used to eliminate stereotypes is to ensure that there is one member of an underrepresented group on every team. However, this use of a "token" member does not seem to alleviate stereotyping behavior, according to the classic book *Men and Women of the Corporation* by Rosabeth Moss Kanter, a business professor at Harvard. In fact, Kanter has found that a token can actually increase stereotyping, especially when the token is negatively stereotyped with respect to the task at hand. Because people look for confirmation rather than disproof (as discussed earlier in the chapter), they will tend to look for evidence that the token member conforms with the stereotype and ignore evidence countering it.

The best way to overcome stereotypes is to ensure that individuals associated with a particular stereotype are a significant percentage of their group, even if that means that other groups will have no representatives of the stereotyped group at all. In fact, the ratio of the stereotyped group to the total size of the group also affects how individuals of the stereotyped group view themselves. This surprising discovery by psychologists Claude Steele and Joshua Aronson indicates that when a member of a negatively stereotyped group with respect to a particular skill (e.g., elderly and memory, males and social judgment, African Americans and language skills) is surrounded by others of a different group, the person's behavior and performance are more likely to reinforce the stereotype. That is, even without being discriminated against, if a situation reminds someone of their affiliation with a stereotyped group, then the automatic cuing of the stereotype leads the person to "choke," a phenomenon Steele and Aronson termed "stereotype threat." Reminders can be as small as pictures of "best performers" that exclude the negatively stereotyped group or being the lone member of a stereotyped group among others not of the group. For example, the only woman in an otherwise all-male engineering group will automatically and unconsciously feel the effects of sexism as compared to when in a balanced-gender or all-female group. While for a few people, stereotypes represent a challenge to triumph over, for the vast majority, the stereotype becomes a threat.

Experiment:

I Am My Group, and That's a Problem

Curbing stereotype threat is particularly important because of its subversive nature: it worsens a person's performance and sadly reinforces the stereotype that the person is trying to avoid. Ph.D. student Roselyn Lee (now a professor at Hope College) and I sought to understand the power of stereotype threat and to find an approach for minimizing its effect by using an online group test-taking situation. We focused on the stereotype that women are less competent at mathematics than men.

In the experiment, female college students took a math test in a computerized setting. The test was composed of twenty questions adopted from the quantitative section of the Graduate Record Examination (GRE). We told participants that they would be taking the test with other test takers remotely logged into the system (in actuality, they were computer confederates). In the test, the participant and "remote" test takers were all represented by avatars (pictorial representations). Using the avatars allowed us to create groups of different gender compositions while keeping the members of the group uniform in every other respect—another advantage of using computers rather than artificially constructed groups. We varied the other test takers' genders: participants were either in a female-majority group (all female or two females and one male) or in a female-minority group (one female—the participant—and two males).

Previous research had suggested that cooperation could shift a person's identification from a cued social group and its associated stereotype to membership in the cooperating group. This is the process described in the Team and Team Building chapter in which interdependence leads to identification. Once teams adopt a shared goal, it becomes more relevant than even well-established differences. By framing the task as cooperative, we could perhaps downplay participants' identification with their gender and thereby eliminate the stereotype threat. Specifically, in the cooperative condition, participants were told that their test-taking group would be compared with other groups of test takers: their goal was to make their group outperform the other groups (i.e., we established interdependence). For comparison, in the competitive condition, the participants were told that their scores would be ranked against the other two test takers and that their goal was to outperform them.

After taking the test, participants filled out a questionnaire about the experience as well as their concerns about stereotypes. For example, participants rated how much they agreed with statements such as: "If I did well on this test, people might consider me as an unusual case given my gender," "If I performed poorly on this test, the researchers of this study might attribute my poor performance to my gender," and "If I didn't do well on this test, it might be viewed as stereotypic of my gender."

➤ Results and Implications

For the typical test-taking situation—competition—being in a minority resulted in stereotype threat: participants in the female-minority groups had significantly lower scores on the math test. Being the only female among males (even though there were only two males) made participants conscious of their gender while their concern for confirming negative stereotypes distracted them from the test and led to lower scores. Participants' self-reported concerns about stereotypes reflected this as well.

What about in the cooperation conditions? Cooperation successfully eliminated the stereotype threat. In the female-minority groups, the cooperation participants did just as well on the test as the participants in the female-majority groups (and much better than the female-minority competition participants): the pressure of representing women was eliminated, replaced with affiliation to the test-taking group. This was confirmed by the questionnaire results, as participants who engaged in cooperation reported significantly lower levels of stereotype-associated concerns.

Stereotype threat is unfortunately easy to evoke, as demonstrated by numerous studies since the original work of Steele and Aronson. Simply place a person who can be associated with a negative stereotype with others outside of the implicated group, and the person's performance can be severely impaired. The person's minority status need not be emphasized or even discussed: a hint as slight as the presence of two other avatars can trigger the effect. A group does not have to be widely stereotyped to be susceptible to stereotype threat. For example, if a certain department is notorious throughout the company (including among the members of the department itself) for being slow, having a single representative from that department on a rush project will likely lead to him or her feeling the tension of expected failure and regrettably living up to the group's negative reputation.

There are other ways besides labels to increase one's perceived expertise in order to become more persuasive, as discussed in my book *Wired for Speech*. For example, people perceive you as more intelligent if you speak more fluidly, that is, with appropriate pausing (you don't talk nonstop or pause abruptly or unnaturally), rapid speech (a Southern drawl makes you seem unintelligent), and avoidance of "uhs," "ums," and Valley Girl speech (the continual use of "like" or "really"). According to education researcher John Follman, even teachers—trained professionals in assessing intelligence—are as likely to judge students' intelligence based on how they speak as on their classroom performance. People also find those who use moderately emotional speech more intelligent than those who use either monotone or highly emotional speech: no emotion suggests thoughtlessness, and extreme emotion suggests irrational thinking. Even without necessarily knowing about this research, advertisers have reached similar conclusions: spokespeople almost always speak quickly and in a smooth, articulate, and expressive manner.

Physical markers can also affect how competent you seem. Attractive people are more persuasive than unattractive people (although extremely attractive people are seen as less intelligent and less persuasive). Juries tend to find better-looking witnesses more believable. Taller people are more persuasive than shorter people. People consider ectomorphs, who are thin, long-limbed, and small-shouldered (Ichabod Crane, Woody Allen) more intelligent and persuasive than either mesomorphs, people with muscular bodies and erect posture (Superman), or endomorphs, people with round physiques and soft bodies (Santa Claus).

As part of your physical appearance, clothes can also influence perceptions of competence. People who wear sloppy clothes or overly informal clothes seem less intelligent than those whose clothes are consistent with or more formal than appropriate. I had to confront this when I started as an assistant professor. Although I'm not color-blind, I would often unintentionally wear outrageously mismatched clothes. The reason was that I developed my sense of style from my father, who could not distinguish colors and frequently wore clothes that clearly (at least to everyone else) clashed.

I realized my poor fashion sense would undermine my credibility and authority, so I asked my friends for help. They developed a brilliant solution: sewing Garanimals in my clothes. Designed to help children select matching garments, the strategy involves taking clothes that go together and sewing the same animal picture into them. When getting dressed, I simply had to check that I had my animal pictures matched. This approach—combined with the consistent use of an iron—made me feel like America's Next Top Model.

After an initial period of suave dressing, however, I found that people were once again looking askance at my clothes. One day, an elderly woman stopped me on the street.

"Excuse me, young man," she said, "but I must tell you that you should not wear colorful stripes with colorful plaids." Instinctively, I turned up the bottom of my shirt and turned down the top of my pants. "Thanks for the advice," I replied, "but I seem to have a lion with a lion." The woman backed away and hurried down the street away from me.

When I later told my friends about the incident, they all started laughing hysterically. It turned out that after gaining my confidence, they had sneaked into my house and resewn the animals so that I would embarrass myself. I subsequently decided that a better solution than relying on my jokester buddies was to buy only solid-colored clothes in a few interchangeable colors.

Perceived expertise can be extremely powerful, even when there is a highly tenuous basis for that judgment. This should have meant that when I was pitching the perfect support agent (described at the start of the chapter), I could not fail. Not only was I a professor (accreditation) with a history as a consultant (experience) and an enormous amount of data to support my conclusions (knowledge), I was known throughout the company as an expert. How could I be any more persuasive?

The problem was that knowledge alone does not guarantee persuasiveness. As mentioned earlier on in the chapter, when someone asks you to do something or believe something, there is another critical consideration: whether the person has your best interests at heart. The decision about whether someone is *worth listening to* is grounded in perceptions of expertise, but the decision of whether someone *should be listened to* is grounded in whether you trust the persuader.

As described by psychologist Daniel Goleman in his book *Emotional Intelligence,* researchers have discovered that having high degrees of social and emotional skills can make you more persuasive. Indeed, Goleman argues that in terms of how well people do in life, "social intelligence" is as important as actual IQ or other more traditional measures of brainpower. For example, in a classic study of star performers at Bell Labs, business consultants Robert E. Kelley and Janet Caplan found that it was the stars' ability to build and utilize their informal network that made them more effective at getting their work done. By cultivating relationships with a wide range of people, the star performers could easily reach out to the right person and get help quickly when a question or problem arose. Superstars appreciated the importance of successful social relationships and of building rich networks of information and people they could trust. In contrast, the researchers observed that when others ran into a technical problem and called or e-mailed several experts, they were less likely to get a response. The importance of informal networks underlines how critical building and sustaining trust is for success in a work environment. Indeed, an effective use of social rules can actually make people seem more knowledgeable and intelligent as well as trustworthy.

When you work with people for an extended time, numerous opportunities arise for you to demonstrate your trustworthiness. And in fact, even brief interactions support trustworthiness because of the principle "familiarity breeds liking and liking breeds trust." I recently completed a study with Maurits Kaptein, a visiting Ph.D. student from Eindhoven University of Technology and Philips Research, and his advisor, Daan van Bel, that explored this idea. We had participants communicate with a stranger via online chat for approximately five minutes. We used the I-Sharing task developed by social psychologist Elizabeth

Pinel and colleagues, which involves exchanging answers to twelve trivial questions such as, "If Colin Powell were a sea creature, he would be a: a) hammerhead shark, b) crab, c) dolphin, or d) octopus."

We then presented participants with scenarios that involved two distinct courses of actions and had either the person they had just communicated with or a different person provide suggestions about which action to choose in the face of the dilemma. We wrote the dilemmas so that they would not have an obvious choice; that is, neither option was clearly more desirable or socially appropriate. For example, one of the dilemmas read as follows:

> Mr. D., a married man with two children, has a steady job that pays him about sixty thousand dollars per year. He can easily afford the necessities of life, but few of the luxuries. His father, who died recently, carried a forty-thousand-dollar life insurance policy. He would like to invest this money in stocks. He is well aware of the secure blue-chip stocks and bonds that would pay approximately 6 percent on his investment. On the other hand, he has heard that the stocks of relatively unknown company X might double their present value if a new product currently in production is favorably received by the buying public. However, if the product is unfavorably received, the stocks would decline in value. Should Mr. D. buy the stock or not?

Did the very minimal online interaction with a person the participant had never met make the person more convincing? Yes. Participants who received suggestions from their previous partner were much more persuaded by their suggestions, although the familiarity was only built on exchanging opinions on twelve silly questions.

While any interaction increases trust, a stronger basis for persuasion is to prove your trustworthiness. One way to do this that societies have developed is the reciprocity test: if you have helped me, then I can trust you. That is, the act of performing a favor demonstrates that someone has your interest in mind, which then makes it likely that her or his suggestions or recommendations should be accepted.

In addition to cultivating trust, reciprocity implicates a responsibility: if someone does a favor for you, you are *obligated* to do something nice in return. This "rule of reciprocity" is one of the few cultural universals: all known human societies dictate, "Do unto others as they have done unto you" (other versions of this phrase exist, but this is the universal form). It's more than a simple prescription; there are very strong social injunctions against and punishments for failing to reciprocate. Indeed, some scientists argue that reciprocity is so ingrained that people do not have to be reminded of their obligations.

My Ph.D. student B. J. Fogg (now a consulting professor at Stanford University) and I wanted to explore just how far you can take reciprocity as a persuasive strategy. Is the rule enacted even when the giver of the favor lacks feelings, is independent of moral considerations, and cannot appreciate the notion of the Golden Rule? Specifically, we investigated whether people would feel that they owed a computer a favor after the computer helped them. In a sense, this is a ludicrous notion: people do not feel that they owe a hammer a favor when it successfully drives in a nail for them, for example. If even a computer can persuade people using reciprocity, then everyone can use reciprocity as a persuasive tactic.

Experiment: ---------------------------

Can a Computer Say, "You Owe Me One"?

Once again, we presented people with a variant of the Desert Survival Situation. In this version, we told participants that they could use a computer to find information about the items they needed to rank. The computer was very helpful: the results from participants' search requests were filled with relevant information. For example, when they asked about flashlights, the computer told them, "The beam from an ordinary flashlight can be seen as far as fifteen miles away on a clear night." The computer also claimed that it had searched many databases to get the best information. That is, not only was the computer being useful, it was also "trying" very hard.

When we were designing the experiment, we figured out how the computer could do a person a favor easily enough but were puzzled how the computer could then ask for a favor in return. If the computer said, "It would make me happy if you did what I asked," or "This computer would feel betrayed if you didn't answer my questions," the artificiality of the computer expressing feelings would likely elicit an artificial response from the participant. Even if the computer simply said, "I need a favor from you," that would be too blatant to really explore the power of reciprocity.

We needed the computer to ask the participant to do something that would obviously benefit the computer without explicitly making the person feel guilty if she or he did not do it. The task also had to be unrelated to the search task because we did not want to implicate issues of expertise. Putting this together, we had the computer ask for the participant's help to create a color palette that best matched human perception. Specifically, the computer told participants that it was developing a new algorithm for translating from RGB (red, green, blue color space) to IHS format (intensity, hue, saturation color space). It said that to develop the algorithm, it needed participants to rank three colors from light to dark. After the participant ranked the three colors, the computer would then present another set of colors to rank. Participants had to do a minimum of five rounds, but they could continue helping the computer as long as they liked. The computer told participants that the more comparisons they did, the better that its algorithms would be as compared to other computers' algorithms.

On the one hand, the computer presented a way to improve itself and indicated that the participant could help. On the other hand, the computer did not reference obligations, morality, reciprocity, or in any other way imply that the participant owed the computer. To see if reciprocity made the computer more persuasive, in addition to recording how many comparisons the participant did and how accurate they were, we gave participants a questionnaire about their feelings related to the experience.

Half of the participants performed the color palette task on the same computer they had used for the research associated with the Desert Survival Situation (giving them the opportunity to reciprocate to the computer that had given them helpful information in the previous task). As a point of comparison, the other half of the participants switched computers after the Desert Survival Situation and received the request for help from a different but identical computer than the one they used in the search task. In this case, if the participants helped the second computer, it would not be because of reciprocity (because the second computer had not done the participant a favor).

➤ Results and Implications

When participants used the same computer again for the second task, they were much more likely to help it as compared to participants who worked with a different computer for each task. Specifically, participants who used the same computer for both tasks completed almost double the number of color evaluations. They made far fewer mistakes than the other participants and reported in their questionnaires that they worked harder on the task. Returning the computer's favor also made participants feel good: they enjoyed doing the work, finding the color-matching task more fun and more interesting.

Thus, the computer was far more effective at enlisting the participant's help if it had first done the participant a favor by being helpful. Participants felt obliged to return the favor, even when the task was unrelated and the recipient was a computer. In contrast, the participants who used two computers had just met the second computer when it asked them for help: they didn't owe it a thing!

When we looked at these results, we realized that there might be another explanation: people could simply prefer working with a familiar computer. To distinguish between familiarity and reciprocity, I needed to see what people would do when reciprocity was not a part of the interaction but familiarity still was. We brought in new participants and had them complete the Desert Survival Situation with a computer that was clearly *unhelpful*. For example, when a participant entered the search term "flashlight," the computer returned, "Small flashlight: easy-to-find yellow Lumilite flashlight is there when you need it. Batteries included." That is, the search engine recognized the search term but provided information that was obviously irrelevant to the participant's survival in the desert. The computer also made it clear that it had searched a very limited number of databases, suggesting that it was either lazy or incompetent. Again, we presented half of the participants with the color palette task on the same computer and half with the task on a second, different computer to see how much they would help.

When examining these conditions, we discovered that participants who worked with the unhelpful computer twice exhibited the flip side of reciprocity: revenge.* That is, when the searching computer was unhelpful and then asked for the participants' help, they made fewer comparisons on it than when asked to do comparisons on a second computer. They also did not try as hard, making more mistakes. In other words, when they worked for a second time with the computer that had failed them, participants saw being less helpful on the second task as a way to seek revenge. Similarly, being more helpful when they worked with a different computer on the second task may have been another form of vengeance. As one might expect, the unhelpful-same-computer participants also disliked the second task and the computer more. This proved that we had really found reciprocity in the first experiment and revenge in the second one rather than familiarity effects.

* * *

---

* In many societies, there is ambiguity about the acceptability of revenge. In the United States, for example, we hear "justice must be served" and "an eye for an eye," on the one hand, and "turn the other cheek," on the other.

In retrospect, I could have used reciprocity to help convince the company I described at the start of the chapter. Instead, I emphasized my independence and made it clear that I was responsible only to the facts. Given that I was oriented to "science" rather than to the team, why should they have felt obligated to take my suggestion? I should have said that my primary reason for doing the research was my desire to make the members of the group look like geniuses by finding them a spectacular character. This would have led the group to feel the need to reciprocate, in this case, by not arbitrarily rejecting my recommendation.

Unfortunately, reciprocity is not a panacea. For example, I value the writing skills of a colleague so much that I wanted him to "owe me one" when I completed the first draft of this book. So whenever he was writing anything, I would immediately offer to read and comment on it. Once I had done this twenty times, I was certain that I had done enough to be able to cash in, likely without even asking. Hence, when he asked me for help on his twenty-first document, I declined. What was the result? He felt that *I* owed *him* a favor because I had failed to come through when *he* needed *me*.*

Experiment: Culture Calamity?

At the time we did this experiment, cross-cultural studies were drawing a great deal of attention, and I had been discussing the possibility of performing comparative studies with Yasunori Katagiri and Yugo Takeuchi of ATR in Japan. This experiment seemed a perfect candidate because few countries have more formal and elaborate rules of reciprocity than Japan.

We worked with our Japanese collaborators to get everything trans-

* Fortunately, Corina Yen is a brilliant writer, making his help (and gratitude) unnecessary.

lated and to make sure that we would run the study the same way in both countries. When we analyzed the resulting data, we were stunned: the two groups of Japanese participants exhibited essentially identical behaviors, with all participants doing approximately the same number of comparisons and with similar accuracy. In a culture in which reciprocity is a core value, we did not see reciprocity!

When we told our Japanese collaborators what we had found, they knew instantly what had happened. "You fail to understand how reciprocity works in Japan," Katagiri told me. "Japanese societies are collectivist; U.S. society is individualist. In the U.S., if you help me, I only owe you, but in Japan, if you help me, I owe your entire family." "That's great," I said morosely. "Last I checked, computers don't procreate, so how can they benefit from the feelings of family?"

"That is where you are wrong," Katagiri replied. "There are two obvious families of computers: PCs and Macs. Perhaps the participants felt that because the two computers were identical, they were from the same family and thus they owed the second computer just as much as the first computer."

To check this hypothesis, we repeated the experiment with two additional conditions: one in which participants received help from a PC and then were asked for help by a Mac, and another in which participants received help from a Mac and then were asked for help by a PC. The results were spectacular: just as my Japanese collaborator expected, when participants worked on a second computer that was from a different "family," they were clearly less willing to help it, regardless of which brand of computer they used first. In addition, participants had much more negative feelings about having to do comparisons on the second, "unrelated" computer.

This study provides insight into one of the key questions about social rules: to what extent do they apply cross-culturally? We discovered that some domains of social rules are the same around the world, but the details may differ. For example, all cultures insist that people follow norms of reciprocity, but to whom one must reciprocate may differ. Similarly, all cultures have rules about differences in male versus female behaviors, but the particular stereotypes and behavioral expectations can vary greatly from culture to culture. By understanding the psychological basis for social rules, we can guess which parts of them will be universal and which parts might be affected by culture.

Experiment: Pushing Reciprocity to the Limits

Researchers have many reactions when they hear about truly surprising results. Some use the discovery to remind themselves that there are always new things to discover. Others simply file it away as material to liven up lectures. The rarest and most interesting researchers say to themselves, "I bet that I can push that study even further." Harvard Business School professor Youngme Moon (who was once my Ph.D. student) is in this third group: having heard about the reciprocity results, she decided that she could plumb the depths of reciprocity.

She started with the observation that the best way to get people to disclose information about themselves is to leverage reciprocity. For example, "I'm an Aries, what's your sign?" gets a more reliable response than "What's your sign?" Similarly, if you say to someone, "I'm from New Jersey; where are you from?" then people feel more comfortable in revealing their own hometown. In her study, Moon was interested in much more personal revelations as well as in having a computer rather than a person ask the questions. Specifically, she decided to see if even the most trivial "admissions" by the computer could persuade participants to divulge highly personal information about themselves.

In her experiment, for half of the participants, the computer solicited answers in a relatively straightforward manner, for example, by saying, "What has been your biggest disappointment in life?" For the other half of the participants, the computer tried to use reciprocity in order to extract more information. In this condition, the computer preceded each interview question by giving some information about itself, for example:

> This computer has been configured to run at an extremely high speed. But 90 percent of computer users don't use applications that require these speeds. So this computer rarely gets used to its full potential. What has been your biggest disappointment in life?

Here are some of the other questions designed to get participants to reciprocate:

> There are times when this computer crashes for reasons that are not apparent to its user. It usually does this at the most inopportune time, causing great inconvenience to the user. What have you done in your life that you feel most guilty about?

> This computer has a very fast processor compared to most other models on the market today. It also has a super-high-density DVD-RW+ drive, which allows it to play movies and CDs in multiple formats. And the hard drive is huge. What characteristics of yourself are you most proud of?

> Computers are built so that they can theoretically last for years and years. However, because newer and faster computers are always coming along, most computers last just a few years before they are dumped by their owners. What are your feelings and attitudes about death?

> This computer's abilities are really limited. For example, this computer can do word processing and spreadsheets, but it cannot do any kind of physical activity, such as play sports or walk down the street. What are some of the things you hate about yourself?

> A few weeks ago, some user came in here and began using this computer to edit some video. No one had ever done this on this computer before. Can you describe the last time that you were sexually aroused?

After reading each question, participants typed their response into the computer, which then gave them their next question. To determine which computer was more persuasive at eliciting personal information (the one that simply asked the questions or the one that first revealed information about itself), Moon measured participants' answers according to three criteria. First, she counted the number of discrete self-disclosures; that is, how many different facts participants revealed about themselves. Second, she counted the number of words that people used in their answers: more words suggested more disclosure. Finally, two independent judges (they did not communicate with each other) and who were blind to experimental condition (they did not know what condition each participant was in) assessed how revealing each answer was on a scale of 1 to 5. Participants also filled out a questionnaire at the end of the study describing their feelings toward and perceptions of the computer.

➤ Results and Implications

The computer that first shared information about itself clearly persuaded participants to disclose more intimate details. For example, when the computer said:

> **You may have noticed that this computer looks just like most of the other PCs on campus. In fact, 90 percent of all computers are the same color, so this computer is not very distinctive in its appearance. What do you dislike about your physical appearance?**

One participant responded with:

> I hate my big hips. I'm a sugar freak, and all that sugar sits on my hips. I also don't like that I have relatively small breasts, but that is nothing compared to the way the size of my hips bothers me. No amount of running or lifting or anything else seems to slim them.

In contrast, in the "nondisclosive" computer condition, when the computer asked, "What do you dislike about your physical appearance?" one participant responded, "I could lose some pudginess and gain more tone, which requires effort."

All three measures confirmed the power of reciprocity to persuade. Participants provided many more facts about themselves and gave longer answers when the "disclosive" computer asked them questions. Participants' answers were also significantly more revealing to the disclosive computer according to the judges. Finally, participants liked the disclosive computer more than the nondisclosive computer.

These experiments on reciprocity highlight a key point about social behavior: the more fundamental and basic the social rule, the less you need to do to get others to follow it. In the previous experiments, an emotionless favor or an essentially nonsensical disclosure encouraged reciprocation. An ingenious experiment conducted by psychologists Ellen Langer, Arthur Blank, and Benzion Chanowitz further illustrates how easily one can leverage a standard rule to persuade people. The experiment investigated how providing a reason can make you persuasive. Their researchers monitored a photocopy machine in a university building, and when they saw a person making a large number of copies, a confederate would walk up to the person at the copy machine and ask to make five copies. For one-third of the participants, the confederate would simply say, "Excuse me. I have five pages. May I use the Xerox machine?" For another third of the participants, the confederate would also give a reason for the request saying, "Excuse me. I have five pages. May I use the Xerox machine *because I'm in a rush*?" (emphasis mine). This reasonable request made a very large difference: while only 60 percent of people allowed the person to go ahead in the first case, 94 percent agreed in the second case. That is, when given a reasonable request, almost everyone complies with the request.

How reasonable did the request have to be? The confederate said to the last group of participants: "Excuse me. I have five pages. May I use the Xerox machine *because I have to make copies*?" (emphasis mine). Did people recognize that the "reason" for the request was utterly redundant and wholly different from being in a rush? Not at all. There was essentially no difference in compliance (93 percent versus 94 percent) between the group who heard a sensible rationale and those who heard a "placebic" or irrelevant rationale. The key point is that your brain rapidly makes evaluations of whether someone is a nice, reasonable person based on triggers as simple as hearing the word "because." If you seem pleasant and give any rationale at all (indicating that you are considerate and polite), the person receiving the request will simply comply. The more one follows the standard structure for the request, the more likely the request will be accepted.

Another example of an "automatic" human response that increases your persuasiveness is one talked about in previous chapters: similarity-attraction. As we have seen, similarity gives rise to trustworthiness. For example, the Personality chapter demonstrated how people instinctively trust those who have similar personalities to themselves, and persuasiveness naturally grows out of trust. Thus, the computer with a similar personality to a participant was more credible and a better salesperson. Similarly, when people experience identification as part of a team, the similarity makes teammates more persuasive: Even similarity that was as simple as a wristband and a border on a computer of the same color was enough to generate persuasiveness.

Trustworthiness (established via familiarity, reciprocity, or similarity) and expertise (supported by labels) clearly increase persuasiveness. In some cases, you can't use both of them. For example, as shown in the Praise and Criticism chapter, criticism can make you seem smarter than does praise, but praise makes one more likeable and trustworthy. And as discussed in the Emotion chapter, highly emotional speech can make one seem trustworthy but can undermine the appearance of competence.

If you have to choose between expertise and trustworthiness, which is more persuasive? For example, if you were going on vacation, would you listen to the recommendations about what to do there from your neighbor or from a native of the city to which you were traveling? Your neighbor is familiar to you and likely more similar to you (trustworthy), but a native of the city is more familiar with the city and more knowledgeable about its attractions (expert). I designed and conducted a study with linguist Nils Dahlbäck's lab in Sweden and our Ph.D. students Seema Swamy and Jenny Alwin, respectively, to investigate the question of which is more persuasive: similarity or expertise.

Experiment:

Is It What You Know, or How Similar You Are?

For the study, we created a Web site for travelers to post recommendations about tourist attractions. The site had information about two cities: New York and Stockholm. We invented all of the places described on the site so that we could use the same destinations for both cities as well as ensure that participants' prior knowledge was irrelevant to the study. We chose a number of different types of tourist attractions and based the descriptions of the attractions on texts from guidebooks and other tourist information. For each topic, we wrote two descriptions that differed only in location details. For example, the restaurant description read:

> When you're in [Stockholm/New York], you've got the chance to experience wonderful food within every category. I prefer Thai food and I will now tell you a little about my favorite restaurant. Take the [Red line for Gamla Stan/7 train to Queens] and stop at Sio Wha. This place really serves sensational Thai food and is so cheap. Order many dishes with all the best from Thai cuisine including seafood, chicken, meat, chilies, curries . . . and along with that, a cool glass of white rice wine. And finish off your meal with Thai pineapple ice cream . . . and I promise that you will most certainly remember this trip with a content smile.

To investigate the effects of similarity-attraction versus expertise, we recruited American participants who were native English speakers and Swedish participants who were born and lived in Sweden but who were also fluent in English (so that they would understand the descriptions). Half of the American participants and half of the Swedish participants worked with the Web site that was about New York and the other half worked with the version that was about Stockholm.

When a person selected one of the attractions, an audio clip of the description would play. All descriptions were presented in English, but for both the New York and the Stockholm versions of the site, half of the American participants and half of the Swedish participants heard the descriptions read by a person with an American accent and the other half with a Swedish accent. After hearing each description, participants answered questions about how they felt about it and the speaker who read it.

➤ Results and Implications

The questionnaire results showed that similarity-attraction did occur: Americans preferred the American voice and Swedes preferred the Swedish voice. When the speaker's accent matched their own, participants liked the voice more and rated it as more trustworthy and competent. Participants' preference for their countrymen extended to them preferring their countrymen's recommendations about foreign cities. Americans were more likely to follow the recommendations about Stockholm made by an American; similarly, Swedes were more persuaded by Swedes' recommendations about New York. Participants also thought that the advice was more valuable in terms of helpfulness, usefulness, and intelligence when it came from someone from their own country.

In addition to participants preferring the descriptions in terms of those subjective variables, they also thought them more valid: Americans rated the information from American speakers as more accurate, and Swedes rated the information from Swedish speakers as more accurate, regardless of the location they were describing. This is particularly remarkable because participants acknowledged in the questionnaire that Swedes would be more familiar with Stockholm and that Americans would know more about New York. Thus, even when expertise is obvious, the bonds of similarity and its resulting trustworthiness are more persuasive.

These results provide more insight into why I failed in my pitch described at the beginning of the chapter. I presented myself as an "objective external consultant" with "independent data" paid for out of the CEO's budget so that I would not have any connection to the group. I emphasized that I performed more and more types of research than any of them had ever done or seen. I was unambiguously *not* similar to them, and that detracted from the persuasiveness of my expertise.

I should have emphasized similarity as well as expertise. For example, rather than talking at length about my personal research team, I could have also reminded them that we were all involved in the initial character selection and in all aspects of the research design (even if it was just perfunctory approval). I should have made sure that we were all asking and answering the same questions and were all sharing in the results. Phrases such as "In all my years as a consultant" highlighted our differences; instead, a "Can you believe how consistent the results are? That's a credit to all of the planning and preparation we put in" would have combined similarity with teamwork and flattery, a tremendously powerful combination. Finally, I could have really engaged the team by giving out T-shirts with the *new* character, making the character part of the team and much harder to reject.

While the study shows that generally trustworthiness is more persuasive than expertise, tailoring which one you emphasize based on how much your audience cares about the issue can be an even more powerful strategy. Specifically, you should try to determine whether your audience is using the *central route* versus the *peripheral route* to process your suggestions, a distinction made by psychologists Richard Petty and John Cacioppo in their now classic Elaboration Likelihood Model. When people use their central route, they are thinking very hard about what they are hearing; that is, the relevant information is at the *center* of their attention. They consider each step of the argument in great detail and try to come up with counterarguments and alternatives. In this case, your expertise trumps your trustworthiness in terms of persuasiveness because your audience is focused on the facts.

There are several cases in which people are more likely to use the central route. For example, people will tend to process arguments centrally when the topic or result of it has great consequence to them. Central processing is also more likely if they will be held personally responsible for the decision emerging from your suggestion. Finally, some people just enjoy thinking hard—called a "need for cognition" by Cacioppo and Petty—and so will likely use the central route regardless of the circumstances.

In contrast, when using the peripheral route of persuasion, people are not thinking hard about the message. That is, the arguments are on the *periphery* of their consciousness. At most, they will quickly judge the validity of suggestions or arguments based on the most obvious and simple criteria and will not search for alternative possibilities. Because trustworthiness eliminates the need to think hard about the specifics of what is being said, when people are using the peripheral route, your trustworthiness will influence them more than your expertise.

People will use the peripheral route when they are unable to think deeply, for example, when they are under time pressure or multitasking. Similarly, if they do not know much about an issue, there is no point wasting time processing arguments, so they will sensibly rely on judgments of the speaker's trustworthiness. Because everyone has limited cognitive resources, using the peripheral route does not reflect a weak or lazy mind: it is valuable to preserve brainpower and deep thinking for important decisions and when it is needed. The peripheral route is also associated with excitement: an action orientation means less thinking and more doing (and if you are trustworthy, doing whatever you say).

Rather than choosing a persuasive strategy based on your audience's current state, you can also frame the conversation to get it oriented toward central versus peripheral processing. In one study, my

Ph.D. student Key Lee and I had participants use a car simulator that provided information via the radio about the environment. Half of the participants were told to think rationally:

> When driving this car, carefully consider the car's performance. Please consider the acceleration, braking, steering responsiveness, comfort, etc. We will be asking you to analyze these aspects of the car after you complete the course. Please take an analytic stance in your assessment of the car.

The other half of the participants were told to think emotionally:

> When driving this car, please focus on your experience of the car. Please consider your emotions, your feelings, your engagement, and your intuitions. We will be asking you to describe these aspects of your feelings after you complete the course. Please take an emotional stance in your feelings about the car.

Half of the rational- and emotional-orientation participants heard messages that supplied lots of facts, appealing to the central route, such as "Global warming has caused temperatures to rise by three to eleven degrees in the past hundred years" and "Two billion kilograms of lead were released into this country's atmosphere by the year 2009." The other half heard parallel messages with personal and emotionally compelling details, appealing to the peripheral route, such as "Temperatures are rising, causing people to be hot, sweaty, unhappy, and leading to heat strokes" and "Children are becoming weak and sick because of high levels of lead in their blood."

As expected, the central route messages persuaded drivers with a rational orientation more, and peripheral route messages persuaded drivers with an emotional orientation more. Thus, even if your audience initially seems oriented more toward central or peripheral thinking, you can influence how they receive your attempts at persuasion through framing your message differently.

* * *

A discussion of what makes people more persuasive would not be complete without identifying behaviors that undermine persuasiveness. For example, in the photocopy experiment, if the confederate had asked in a hesitating or guilty manner, the listener would probably have picked that up, grown apprehensive, and scrutinized the request more carefully.

As discussed in the previous chapter, a common marker of dishonesty is when your emotions don't seem genuine, which often occurs when the signs of how you are feeling contradict each other (physical signs of your emotion are naturally consistent). People often inadvertently manifest inconsistent verbal and nonverbal cues when they are trying very hard to be persuasive and have scripted deliveries. For example, sometimes speakers will plan to place a key gesture at a certain point in a presentation. During the speech, though, they might forget the correct moment and end up gesticulating at a different point, thereby mismatching verbal and nonverbal behavior. President George H. W. Bush was well known for these types of awkward and ill-timed gestures. Although he was arguably one of our most intelligent presidents, he would often appear incompetent (in his case, it was likely due to poor oratory skills rather than to a lack of preparation).

Mixed signals make you seem deceptive and untrustworthy, which makes people wary of believing you. I therefore wanted to investigate further to see if people's negative reactions to inconsistency are so automatic that they occur even when the inconsistency is clearly not based in deception. For example, what if your accent does not "match" your race? While people are considerably more mobile now than ever before, our physical characteristics still tend to be very tightly coupled with our accent, language, and culture. It seems natural for an Asian person to speak English with a Korean accent and for a Caucasian person to speak English with an Australian accent. So how would an American react to someone who looks Asian but was born and raised in Australia so that he speaks with an Australian accent? Clearly the inconsistency is not intentionally deceptive or even within the person's control. Do people distinguish between the uncommon and the suspicious?

Experiment:

But I Didn't Choose to Be Inconsistent . . .

To investigate this question, Stanford undergraduate Sheba Najmi (now an interaction designer) and I used a business transaction situation in which the salesperson was a computer agent whose looks and speech reflected either just one ethnicity or two different ones. The context was an online e-commerce site that sold four different products. Each product was depicted via a photograph next to which was a photograph of the salesperson for the Web site. For half of the participants, the salesperson was a Caucasian male; for the other half, the salesperson was an Asian male (both were approximately the same age, the same height and weight, and the same level of attractiveness).

When participants arrived on a product page, they would hear an audio clip of the salesperson reading the product description. For half of the participants, the salesperson "sounded like they looked": the Caucasian salesperson had an Australian accent and used a few typical Australian words (e.g., "G'day, mate"), while the Asian salesperson had a Korean accent and used a few typical Korean words (e.g., "Anyong-haseyo," which means "hello"). For the other half of the participants, the Caucasian salesperson spoke with the Korean accent and words, while the Asian salesperson spoke with the Australian accent and words.* In all other respects, the sales pitches were identical.

* The selection of ethnic groups was inspired by two acquaintances, although we didn't use them for this study. One person was of Asian ancestry who grew up in Australia. He had a thick Australian accent and used Australian terminology. The other, who was Caucasian and grew up in Korea, had a Korean accent and used

We carefully selected products that would not benefit from any "stereotypical expertise": a backpack, a bicycle, an inflatable couch, and a desk lamp. This ensured that the dimension of inconsistency did not relate to what the salesperson was trying to sell.

After hearing each product description, participants answered questions about how much they liked the product and how credible the description of it seemed. We could then see which "salesperson" was most effective at persuading people. At the end of the study, the participants rated the agent overall.

➤ Results and Implications

Even though the consistency between physical ethnicity and accent was out of the salespeople's control, participants found the consistent salespeople more convincing. Specifically, participants liked the products more when presented by an agent who either looked and sounded Korean or looked and sounded Australian. Participants also found the matched agents' product descriptions to be more credible, even though the descriptions were identical in all conditions. Finally, participants found the matched sales agents overall to be more credible, intelligent, and honest. Thus, even the most minor and fully explicable inconsistency strongly undermines persuasion.

While there are many advantages of being persuasive, perhaps the best one is that successful persuasion creates more successful persuasion! Every time you convince someone to do something for you, the more likely she or he is to do something when you ask the next time as well. This is illustrated in a classic *Arabian Nights* story (verified by research starting with psychologists Jonathan Freedman and Scott Fraser's study concerning "the foot-in-the-door" effect). A camel asks a man in

---

Korean words. Both of them told me that they commonly experience the negative effects we found in the study.

a tent to allow it to put its nose inside the tent because of the cold; after all, the camel points out, its nose will not take up much room. After the man lets the camel put its nose in, the camel asks permission for its neck to come in; after all, it won't take much room. Eventually, the camel is in the tent and the man is out in the cold. The moral is that if you can get people to do a small favor for you, it is much easier to persuade them to do a subsequent favor because your next request will seem like a natural extension of their original compliance.

- Your persuasiveness comes down to whether people perceive you as expert (are you *worth* listening to?) and trustworthy (*should* you be listened to?).

- Being labeled an "expert" or a "specialist" grants you all the persuasive power that actual experts have.

- Stereotypes remain powerful because they can undermine (or grant you) expertise. Keep minorities well represented in a group to avoid stereotype threat and increased stereotyping.

- Reciprocity can gain compliance and trust: do someone a favor, and no matter how small, your next request will likely be accepted.

- Trustworthiness is generally more persuasive than expertise. You can also tailor which you emphasize based on whether your audience is using its central or peripheral route for cognition.

- Inconsistency, whether under your control or not, makes you less persuasive.

Epilogue

At the beginning of this book, I promised to show you that the social world is much less complicated than it seems. Now that you have been a vicarious participant in almost thirty of my experiments, I hope that you see how true this is.

Paraphrasing the brilliant interface designer Karen Fries, "Finding the simple is complicated." I've had to put participants in my experiments through many struggles and travails: answering difficult math problems amid the pressure of stereotypically superior competitors (in the form of avatars), dealing with a nagging passenger and frustrating roads on a drive (in a driving simulator that talks), enduring false praise and criticism (from a game-playing computer), facing insults from a virtual twin (via watching a screen), and even wearing a blue wristband (when working with a green computer). The efforts were not in vain: my experiments have exposed the most powerful social principles for improving your day-to-day interactions and relationships. For example, knowing the utility of similarity-attraction, you can successfully maintain teams, work with diverse personalities, calm the frustrated, cheer up the sad, and persuade the dubious. People's feelings and behaviors related to giving and receiving evaluations, getting along with different personalities, building teams, managing emotions, and persuading others are grounded in just a few fundamental principles, such as consistency and hedonic asymmetry.

I've uncovered many of these findings through my discovery that people treat computers and other interactive technologies like actual people. Watching people work with computers in social situations lets me strip away complexity and get to the fundamental truths of everyone's interactions.

The rules I have collected aren't just simple: they are powerful. It's hard to imagine anyone who would have a tougher time mastering the social graces than a box of plastic and silicon. Nonetheless, when we apply social insights and strategies to their designs, computers, cars, and other technologies have richer and more compelling interactions with people, becoming effective salespeople, teachers, passengers, persuaders, evaluators, and confidants. Thus, the resulting rules (because they are effective when a mere computer acts in the place of a person) can be readily and effectively applied by anyone.

With the insights and principles presented here, you can become more socially successful and effective. Trust in simplicity and confidently use the science and applications in this book. And if you have any doubts, just remember: *If a computer can do it, so can you.*

Acknowledgments

It is said that the key to business success is to find a friend and a mentor. A scientist also needs an outstanding lab group. I have been extremely lucky to assemble a group of students who are not only extraordinarily intelligent and passionate about research but are also extremely kind and caring. They have contributed to the book through their commentary and emotional support as well as through their research. Although I cannot thank everyone who has worked in my lab over the past twenty-four years, I do want to note those who have been a part of the lab while I have been writing the book: Dean Eckles, Nicole Fernandez, Victoria Groom, Bhavna Hariharan, Helen Harris, Joris Janssen, Jae Min John, Yeon Joo, Malte Jung, Maurits Kaptein, Jeamin Koo, J. Roselyn Lee, Key Lee, Katherine Murray, Michael Nowak, Paloma Ochi, Eyal Ophir, Shailendra Rao, Erica Robles, Neeraj Sonalkar, Piya Sorcar, Abhay Sukumaran, Leila Takayama, and Qianying Wang. I am also very appreciative of industry collaborators who have been part of my lab group during the writing of this book, including Tico Ballagas, Kimi Iwamura, Jofish Kaye, Norimasa Kishi, Brian Lathrop, Takeshi Mitamura, Jo Ann Sison, Mirjana Spasojevic, and Takashi Sunda.

Among my colleagues in the Communication Department, Jeremy Bailenson has been extremely influential in my thinking about various issues discussed in this book. Jeremy, Fred Turner, and Byron Reeves were "book-writing buddies": dear friends, great thinkers, and fellow sufferers in the process of producing a manuscript. Many other Stanford colleagues, most notably Sven Beiker, Keith Devlin, Chris Gerdes, James Gross, Chuck House, Lauri Kanerva, Barbara Karanian, Martha Russell, Michael Shanks, Syed Shariq, Sebastian Thrun, and Anthony Wagner, have also been significant contributors to this book.

Although I study social rules, it has been apparent to my marvelous support staff that I do not understand Stanford's rules. Many thanks to my personal administrators, Erina DuBois and Regina Grimsby, and the other staff who have been invaluable: Kristin Burns, Mark DeZutti, Susie Ementon, Barbara Kataoka, Lillian Kamal, Martha Langill, Susie Ng, Helen Roque, Mark Sauer, and Katrin Wheeler.

Beyond the Stanford people listed above, more people than I can count have made significant contributions to improving this book. Mark and Bonita Thompson have taught me an enormous amount about book-writing and the publishing process. Jason Riggs figured out an ingenious approach for presenting the experiments in this book. Sebastian Yen did a marvelous job designing and creating the illustrations when there was no time to spare.

Joe Bankman, Gabe Bankman-Fried, Sam Bankman-Fried, Barbara Fried, and Matthew Nass have been involved in every up and down of the manuscript and all provided numerous valuable suggestions. Others who have provided very helpful suggestions include David Arfin, Scott Brave, Madeline Chaleff, Sarah Delson, Ken Favaro, David Garfinkle, Larry Kramer, Daniel Kreiss, Sylvia Kurpanek, Kathy Lung, Glenn Magid, Julie Munger, John Parsley, Trevor Pinch, Kathryn Segovia, Bonnie Shipper, Ron Shipper, Justin Solomon, Caroline Suen, Barr Taylor, and David Voelker.

A continuing inspiration for my research has been the wonderful staff and students it has been my pleasure to work with as a professor and especially as a Resident Fellow ("Dorm Dad") at Otero, an all-freshman dorm on the Stanford campus. Watching my Oteroans grow into young women and men has made me appreciate how difficult and powerful it is to learn the rules of social interaction. I especially want to acknowledge the amazing staff that I have worked with over the past four years—they know more about teaching social rules than almost anyone else I know.

My agent, Carol Roth, encouraged me to write this book, introduced me to all aspects of the world of trade publishing, and saw me through almost the entire process until she passed away just before completion of this manuscript: she will certainly be missed.

The people at Penguin and Current have done outstanding work and were a pleasure to work with. David Moldawer has been a superb editor, making enormous improvements to the manuscript and rapidly and frankly answering numerous e-mail inquiries. Jacquelynn Burke, Christy D'Agostini, and Elizabeth Shreve have been wonderful and creative publicists. Adrian Zackheim, my publisher, grasped the concept of the book immediately and has been a tremendous supporter. Many thanks also to Emily Angell (editorial assistant), Faren Bachelis (copy editor), Sabrina Bowers (book designer), Barbara Campo (production editor), Jaya Miceli (cover designer), and Christoph Niemann (cover art); their work has been uniformly excellent.

My extended family and my friends have been marvelous through the writing process. Too many of them have passed away while I was writing this book. Having already lost both parents and my brother, each loss weighed very heavily on me.

Matthew Nass is a fabulous son and a wonderful colleague and collaborator. Matthew's unwillingness to become depressed in the face of a series of horrible circumstances that no teenager should have to deal with has been inspirational. He kindly allowed me to use the numerous "Matthew stories" scattered through the text and the footnotes even though he vehemently denies their accuracy. Finally, he has given me more joy every day since his birth than anyone could possibly ask or deserve.

Corina Yen has been indispensable and remarkable. She is a marvelous writer with a great ear. She blends creativity with style and patience. Although she is trained as a designer and an engineer, she immediately grasped a very wide range of social science theories, methods, and applications. Her incredible ability to capture the feel and tone of the book was remarkable. Our disagreements over what to say and how to say it were ferocious, impassioned, and ego free: we just wanted to get the right answer. Most important, she is a fabulous person and a great friend.

➤ Corina's Personal Acknowledgments

I want to sincerely thank Cliff for the opportunity to write this book with him and for the resulting incredible journey. I have truly valued our frank and lively discussions that, along with his warmth and humor, have made this such a great collaboration. It has been amazing and a pleasure to work with such a remarkable scientist, a natural storyteller, and an exceptional mentor as Cliff.

Thank you to my family for all their love and support: my father, David, for his guidance and pushing me to be better; my mother, Julie, for her empowering, unwavering belief in me; my sister Jacqueline, for her spirit that inspires me and her astute and always honest opinions; and my baby brother, Sebastian, the kindest and sweetest soul and the coolest kid I know.

I am also grateful for my friends and colleagues. In particular, James, for always being there to listen, and Chris, for his steadfast encouragement and support. Lastly, thank you to all the Bay Area cafés that supplied espresso and WiFi and routinely accommodated me and my MacBook throughout the writing of this book.

➤ Final Words

The Bibliography acknowledges many of those not mentioned above; we truly appreciate these scholars' contributions to our thinking and to the book.

It is sadly inevitable that we have unintentionally omitted at least one person from the lists above; we apologize in advance.

If we have made any errors in this book, we will do our best to correct them on the book's Web site.

Bibliography

➤ Introduction

Aronson, E. (1995). *The social animal* (7th ed.). New York: Freeman.

Fogg, B. J., and Nass, C. (1997). Silicon sycophants: The effects of computers that flatter. *International Journal of Human-Computer Studies, 46*(5), 551–561.

Groom, V., Takayama, L., Ochi, P., and Nass, C. (2009). I am my robot: The impact of robot-building and robot form on operators. In *Proceedings of the human-robot interaction conference: HRI 2009* (pp. 31–36). San Diego, CA: ACM.

Lee, J. R., Nass, C., Brave, S., Morishima, Y., Nakajima, H., and Yamada, R. (2007). The case for caring co-learners: The effects of a computer-mediated co-learner agent on trust and learning. *Journal of Communication,* 57(2), 183–204.

Nass, C., and Brave, S. B. (2005). *Wired for speech: How voice activates and enhances the human-computer relationship*. Cambridge, MA: MIT Press.

Nass, C., Moon, Y., and Carney, P. (1999). Are people polite to computers? Responses to computer-based interviewing systems. *Journal of Applied Social Psychology*, 29(5), 1093-1110.

Nass, C., Moon, Y., and Green, N. (1997). Are computers gender-neutral? Gender-stereotypic responses to computers with voices. *Journal of Applied Social Psychology*, 27(10), 864–876.

Ochi, P., Rao, S., Takayama, L., and Nass, C. (2010). Predictors of user perceptions of Web recommender systems: How the basis for generating experience and search product recommendations affects user responses. *International Journal of Human-Computer Studies*, 68(8), 472–482.

Ophir, E., Nass, C., and Wagner, A. (2009). Cognitive control in media multitaskers. *Proceedings of the National Academy of Sciences, 106*(33), 15583–15587.

Reeves, B., and Nass, C. (1996). *The media equation: How people treat computers, television, and new media like real people and places.* New York: Cambridge University Press.

Takayama, L., and Nass, C. (2008). Assessing the effectiveness of interactive media in improving drowsy driver safety. *Human Factors: The Journal of Human Factors and Ergonomics, 50*(5), 772–781.

Zajonc, R. B. (1965). Social facilitation. *Science, 149*, 269–274.

➤ Chapter 1

Amabile, T. M. (1983). Brilliant but cruel: Perceptions of negative evaluators. *Journal of Personality and Social Psychology, 19*, 146–156.

Amabile, T. M., and Glazebrook, A. H. (1981). A negativity bias in interpersonal evaluation. *Journal of Personality and Social Psychology, 18*, 1–22.

Delis, D. C., Kaplan, E., and Kramer, J. H. (2001). *Delis-Kaplan Executive Function System.* San Antonio, TX: The Psychological Corporation.

Dweck, C. S. (2007). *Mindset: The new psychology of success.* New York: Ballantine Books.

Fogg, B. J., and Nass, C. (1997). Silicon sycophants: The effects of computers that flatter. *International Journal of Human-Computer Studies, 46*(5), 551–561.

Frijda, N. H. (1988). The laws of emotion. *American Psychologist, 43*, 349–358.

Joo, Y., and Nass, C. (unpublished). Make it easy, but tell them it's hard: Framing drivers' perceptions of the road. Stanford University.

Kipling, R. (2010). "If." In *The works of Rudyard Kipling.* London: Nabu Press.

Lezak, M. D., Howieson, D. B., and Loring, D. W. (2004). *Neuropsychological assessment* (4th ed.). New York: Oxford University Press.

Mason, L., and Nass, C. (1989). Partisan and non-partisan readers' perceptions of political enemies and newspaper bias. *Journalism Quarterly*, 66, 564–570.

Mezulis, A. H., Abramson, L. Y., Hyde, J. S., and Hankin, B. L. (2004). Is there a universal positivity bias in attributions? A meta-analytic review of individual, developmental, and cultural differences in the self-serving attributional bias. *Psychological Bulletin, 130*(5), 711–747.

Mutz, D. C., and Reeves, B. (2005). The new videomalaise: Effects of televised incivility on political trust. *American Political Science Review*, 99(1), 1–15.

Nass, C., and Brave, S. B. (2005). *Wired for speech: How voice activates and enhances the human-computer relationship*. Cambridge, MA: MIT Press.

Nass, C., and Steuer, J. (1993). Voices, boxes, and sources of messages: Computers and social actors. *Human Communication Research*, *19*(4), 504–527.

Rao, S., and Nass, C. (unpublished). Relieving frustrations via mindsets. Stanford University.

Reeves, B., and Nass, C. (1996). *The media equation: How people treat computers, television, and new media like real people and places*. New York: Cambridge University Press.

Schrauf, R. W., and Sanchez, J. (2004). The preponderance of negative emotion words in the emotion lexicon: A cross-generational and cross-linguistic study. *Journal of Multilingual and Multicultural Development*, *25*(2), 266–284.

Schudson, M. (1981). *Discovering the news: A social history of American newspapers*. New York: Basic Books.

Thornton, R. J. (2005). *Lexicon of intentionally ambiguous recommendations* (2nd ed.). New York: Barnes and Noble Books.

➤ Chapter 2

Ambady, N., and Rosenthal, R. (1992). Thin slices of behavior as predictors of interpersonal consequences: A meta-analysis. *Psychological Bulletin*, *111*, 256–274.

Aronson, E., and Linder, D. (1965). Gain and loss of esteem as determinants of interpersonal attractiveness. *Journal of Experimental Social Psychology*, *1*, 156–171.

Asch, S. E. (1946). Forming impressions of personality. *Journal of Abnormal and Social Psychology*, *41*, 1230–1240.

Cantor, N., and Mischel, W. (1979). Prototypes in person perception. *Advances in Experimental Social Psychology*, *12*, 3–52.

"Desert Survival Situation" is trademarked by and was used by permission of Human Synergistics.

Graves, J. R., and Robinson, J. (1976). Proxemic behavior as a function of inconsistent verbal and nonverbal messages. *Journal of Counseling Psychology*, *23*, 333–338.

Isbister, K. (2006). *Better game characters by design: A psychological approach*. San Francisco: Morgan Kaufmann.

Isbister, K., and Nass, C. (2000). Consistency of personality in interactive characters: Verbal cues, non-verbal cues, and user characteristics. *International Journal of Human-Computer Studies*, *53*(1), 251–267.

Kiesler, D. J. (1983). The 1982 interpersonal circle: A taxonomy for complementarity in human transactions. *Psychological Review, 90*, 185–214.

Lafferty, J. C., and Eady, P. M. (1974). *The desert survival problem.* Plymouth, MI: Experimental Learning Methods.

Leary, T. (1957). *Interpersonal diagnosis of personality.* New York: Ronald Press.

Moon, Y. (1998). *When the computer is the "salesperson": Computer responses to computer "personalities" in interactive marketing situations* (Working Paper). Harvard University.

Moon, Y., and Nass, C. (1996). How "real" are computer personalities? Psychological responses to personality types in human-computer interaction. *Communication Research*, 23(6), 651–674.

———. (1998). Are computers scapegoats? Attributions of responsibility in human-computer interaction. *International Journal of Human-Computer Studies*, 49(1), 79–94.

Nass, C., and Brave, S. B. (2005). *Wired for speech: How voice activates and enhances the human-computer relationship.* Cambridge, MA: MIT Press.

Nass, C., and Lee, K. (2001). Does computer-synthesized speech manifest personality? Experimental tests of recognition, similarity-attraction, and consistency-attraction. *Journal of Experimental Psychology: Applied*, 7(3), 171–181.

Nass, C., and Moon, Y. (2000). Machines and mindlessness: Social responses to computers. *Journal of Social Issues*, 56(1), 81–103.

Nass, C., Moon, Y., Fogg, B. J., Reeves, B., and Dryer, D. C. (1995). Can computer personalities be human personalities? *International Journal of Human-Computer Studies*, 43(2), 223–239.

Reeves, B., and Nass, C. (1996). *The media equation: How people treat computers, television, and new media like real people and places.* New York: Cambridge University Press.

Roberts, B. W., and del Vecchio, W. F. (2000). The rank-order consistency of personality traits from childhood to old age: A quantitative review of longitudinal studies. *Psychological Bulletin*, 126(1), 3–25.

Wiggins, J. S., and Broughton, R. (1985). The interpersonal circle: A structural model for the integration of personality research. *Perspectives in Personality, 1*, 1–47.

➤ Chapter 3

Aronson, E., and Mills, J. (1959). The effect of severity of initiation on liking for a group. *Journal of Abnormal and Social Psychology*, 59, 177–181.

Axelrod, R. (1984). *The evolution of cooperation.* New York: Basic Books.

Axelrod, R., and Hamilton, W. D. (1981). The evolution of cooperation. *Science, 211*, 1390–1396.

Beniger, J. R. (1987). Personalization of mass media and the growth of pseudo-community. *Communication Research, 14*(3), 352–371.

Cialdini, R. B. (2007). *Influence: The psychology of persuasion*. New York: Collins.

Cialdini, R. B., Borden, R. J., Thorne, A., Walker, M. R., Freeman, S., and Sloan, L. R. (1976). Basking in reflected glory: Three (football) field studies. *Journal of Personality and Social Psychology*, 34(3), 366–375.

Dawkins, R. (1989). *The selfish gene* (new ed.). Oxford: Oxford University Press.

"Desert Survival Situation" is trademarked by and was used by permission of Human Synergistics.

Durkheim, E. (1947). *The division of labor in society* (G. Simpson, trans.). New York: Free Press.

Gawande, A. (2009). *The checklist manifesto: How to get things right*. New York: Metropolitan Books.

Halberstam, D. (1993). *The best and the brightest*. New York: Ballantine.

Janis, I. L. (1971). Groupthink among policy makers. In N. Sanford and C. Comstock (eds.), *Sanctions for evil* (pp. 71–89). San Francisco: Jossey-Bass.

Jones, J. T., Pelham, B. W., Carvallo, M., and Mirenberg, M. C. (2004). How do I love thee? Let me count the Js: Implicit egotism and interpersonal attraction. *Journal of Personality and Social Psychology*, 87(5), 665–683.

Jourdane, M. (2005). *The struggle for the health and legal protection of farm workers: El cortito*. Houston, TX: Arte Publico.

Katzenbach, J. R., and Smith, D. K. (2003). *The wisdom of teams: Creating the high-performance organization*. New York: Collins.

Kennedy, S. (1990). *The Klan unmasked*. Gainesville, FL: University Press of Florida.

Lafferty, J. C., and Eady, P. M. (1974). *The desert survival problem*. Plymouth, MI: Experimental Learning Methods.

Lee, E.-J., Nass, C., and Brave, S. (2000). Can computer-generated speech have gender? An experimental test of gender stereotypes. In *CHI '00 extended abstracts on human factors in computing systems* (pp. 289–290). The Hague, Netherlands: ACM.

Levitt, S., and Dubner, S. J. (2005). *Freakonomics: A rogue economist explores the hidden side of everything*. New York: William Morrow.

Mabogunje, A. (1997). *Measuring Mechanical Design Process Performance: A Question Based Approach*. Unpublished dissertation, Stanford University, Stanford, CA.

Mabogunje, A., and Leifer, L. (1997). *Noun phrases as surrogates for measuring early phases of the mechanical design process.* Paper presented at the 9th International Conference on Design Theory and Methodology, New York.

McLean, B., and Elkind, P. (2004). *The smartest guys in the room: The amazing rise and scandalous fall of Enron*. New York: Penguin Portfolio.

Merton, R. K. (1946). *Mass persuasion: The social psychology of a war bond drive*. New York: Harper.

Meyerson, D., Weick, K. E., and Kramer, R. M. (1996). Swift trust and temporary systems. In R. M. Kramer and T. R. Tyler (eds.), *Trust in organizations: Frontiers of theory and research* (pp. 166–195). Thousand Oaks, CA: Sage.

Nass, C., Fogg, B. J., and Moon, Y. (1996). Can computers be teammates? *International Journal of Human-Computer Studies, 45*(6), 669–678.

Nass, C., Kim, E-Y., and Lee, E-J. (1998). When your face is the interface: An experimental comparison of interacting with one's own face or someone else's face. In *Proceedings of the SIGCHI conference on human factors in computing systems* (pp. 148–154). Los Angeles, CA: ACM.

Rogers Commission. (1986). *Report of the Presidential Commission on the Space Shuttle Challenger Accident.* Retrieved from http://history.nasa.gov/rogersrep/genindex.htm.

Stoner, J. A. F. (1968). Risky and cautious shifts in group decisions: The influence of widely held values. *Journal of Experimental Social Psychology*, 4, 442–459.

Williams, K. D. (2007). Ostracism. *Annual Review of Psychology*, 58, 425–452.

➤ Chapter 4

Ariely, D., and Loewenstein, G. (2005). The heat of the moment: The effect of sexual arousal on sexual decision making. *Journal of Behavioral Decision Making, 19*, 87–98.

Clouse, R. W., and Spurgeon, K. L. (1995). Corporate analysis of humor. *Psychology: A Journal of Human Behavior*, 32(3-4), 1–24.

"Desert Survival Situation" is trademarked by and was used by permission of Human Synergistics.

Duncan, W. J., Smeltzer, L. R., and Leap, T. L. (1990). Humor and work: Applications of joking behavior to management. *Journal of Management and Information Systems, 16*(2), 255–278.

Duncker, K. (1945). On problem-solving. *Psychological Monographs, 58* (whole no. 5).

Dutton, D. G., and Aron, A. P. (1974). Some evidence for heightened sexual attraction under conditions of high anxiety. *Journal of Personality and Social Psychology*, 30, 510–517.

Ferstl, E. C., Rinck, M., and von Cramon, D. Y. (2005). Emotional and temporal aspects of situation model processing during text comprehension: An event-related fMRI study. *Journal of Cognitive Neuroscience, 17*(5), 724–739.

Freedman, J. L., and Fraser, S. C. (1966). Compliance without pressure: The foot-in-the-door technique. *Journal of Personality and Social Psychology, 4*(2), 195–202.

Goleman, D. (1995). *Emotional intelligence: Why it can matter more than IQ.* New York: Bantam Books.

Gray, J. (1993). *Men are from Mars, women are from Venus: A practical guide for improving communication and getting what you want in your relationships.* New York: HarperCollins.

Gross, J. J. (1999). Emotion and emotion regulation. In L. A. Pervin and O. P. John (eds.), *Handbook of personality: Theory and research* (2nd ed., pp. 525–552). New York: Guilford Press.

Harvey, P. (1990). *Introduction to Buddhism.* New York: Cambridge University Press.

Hatfield, E., Cacioppo, J. T., and Rapson, R. L. (1994). *Emotional contagion.* New York: Cambridge University Press.

Heller-Roazen, D. E., and Mahdi, M. E. (2008). *The Arabian nights (Norton critical editions)* (H. Haddawy, trans.). New York: W. W. Norton.

Hovland, C. I., and Weiss, W. (1951). The influence of source credibility on communication effectiveness. *Public Opinion Quarterly, 15*(4), 635–650.

Isen, A. M., Rosenzweig, A. S., and Young, M. J. (1991). The influence of positive affect on clinical problem solving. *Medical Decision Making, 11*(3), 221–227.

Joo, Y. K., and Nass, C. (unpublished). Effects of valence and arousal on blame attribution in a car simulator. Stanford University.

Kaptein, M., Nass, C., and van Bel, D. (unpublished). Better the devil you know than the devil you don't. Stanford University.

Katagiri, Y., Nass, C., and Takeuchi, Y. (2001). Cross-cultural studies of the computers are social actors paradigm: The case of reciprocity. In M. J. Smith, G. Salvendy, D. Harris and R. Koubek (eds.), *Usability evaluation and interface design: Cognitive engineering, intelligent agents, and virtual reality.* (pp. 1558–1562). Mahwah, NJ: Lawrence Erlbaum Associates.

Keinan, G. (1987). Decision making under stress: Scanning of alternatives under controllable and uncontrollable threats. *Journal of Personality and Social Psychology, 52*(3), 639–644.

Klein, J., Moon, Y., and Picard, R. (2002). This computer responds to user frustration: theory, design and results. *Interacting with Computers, 14*, 119–140.

Lafferty, J. C., and Eady, P. M. (1974). *The desert survival problem*. Plymouth, MI: Experimental Learning Methods.

Lang, P. J. (1995). The emotion probe: Studies of motivation and attention. *American Psychologist, 50*(5), 372–385.

Langer, E., Blank, A., and Chanowitz, B. (1978). The mindlessness of ostensibly thoughtful action: The role of placebic information in interpersonal interaction. *Journal of Personality and Social Psychology, 36*, 635–642.

Lazarus, R. S., and Alfert, E. (1964). Short-circuiting of threat by experimentally altering cognitive appraisal. *Journal of Abnormal and Social Psychology, 69*, 195–205.

Marshall, P. (director). (1992). *A league of their own* [Motion picture]. United States: Sony Pictures.

Moon, Y. (1990). Intimate exchanges: Using computers to elicit self-disclosure from consumers. *The Journal of Consumer Research*, 26(4), 323–339.

Morkes, J., Kernal, H. K., and Nass, C. (2000). Effects of humor in task-oriented human-computer interaction and computer-mediated communication: A direct test of SRCT theory. *Human-Computer Interaction, 14*(4), 395–435.

Nass, C., Foehr, U., Brave, S., and Somoza, M. (November 2-4, 2001). *The effects of emotion of voice in synthesized and recorded speech.* Paper presented at the Emotional and Intelligent II: The Tangled Knot of Social Cognition, North Falmouth, MA.

Nass, C., Harris, H., David, K., Diaz, D., Roth, S., and Sullivan, B. (unpublished). Car-tharsis?: A study of emotion regulation for frustrated drivers. Stanford University.

Nass, C., Jonsson, I.-M., Harris, H., Reaves, B., Endo, J., Brave, S., and Takayama, L. (2005). Improving automotive safety by pairing driver emotion and car voice emotion. In *CHI '05 extended abstracts on human factors in computing systems* (pp. 1973-1976). Portland, OR: ACM.

Nass, C., and Najmi, S. (unpublished). Race vs. culture in computer-based agents and users: Implications for internationalizing websites. Stanford University.

Newhagen, J. E., and Reeves, B. (1992). The evening's bad news: Effects of compelling negative television news images on memory. *Journal of Communication*, 42(2), 25–41.

Newport, E. (1990). Maturational constraints on language learning. *Cognitive Science, 14*, 11–28.

Osgood, C. E., Suci, G. J., and Tannenbaum, P. H. (1967). *The measurement of meaning*. Champaign, IL: University of Illinois Press.

Rao, S., Bailenson, J., and Nass, C. (in press). Making personalization feel more personal: A four-step cycle for advancing the user experience of personalized recommenders and adaptive systems. In M. Eastin, T. Daugherty, and N. M. Burns (eds.), *Handbook of research on digital media and advertising: User-generated content consumption*. New York: IGI Global.

Richards, J. M., and Gross, J. J. (2000). Emotion regulation and memory: The cognitive costs of keeping one's cool. *Journal of Personality and Social Psychology*, 79, 410–424.

Schachter, S. (1959). *The psychology of affiliation*. Stanford, CA: Stanford University Press.

Shakespeare, W. S. (1975). *The complete works of William Shakespeare*. New York: Gramercy Books.

Soussignan, R. (2002). Duchenne smile, emotional experience, and automatic reactivity: A test of the facial feedback hypothesis. *Emotion*, 2(1), 52–74.

Speisman, J. C., Lazarus, R. S., Morkdoef, A., and Davison, L. (1964). Experimental reduction of stress based on ego-defense theory. *Journal of Abnormal and Social Psychology*, 68, 367–380.

Steinberg, L. (2007). Risk taking in adolescence: New perspectives from brain and behavioral science. *Current Directions in Psychological Science*, 16(2), 55–59.

Sutton, Robert I. (2007). *The no asshole rule: Building a civilized workplace and surviving one that isn't*. New York: Business Plus.

Tsai, J. L., Knutson, B., and Fung, H. H. (2006). Cultural variation in affect valuation. *Journal of Personality and Social Psychology*, 90(2), 288–307.

Warren, A.-M. (2010). Words and emotions. From http://ezinearticles.com/?Words-and-Emotions&id=1495344

Wijermans, N., Jorna, R. E., Jager, W., and Van Vliet, T. (2008). *Modelling crowd dynamics: Influence factors related to the probability of a riot*. Paper presented at the 2nd World Congress on Social Simulation, Fairfax, VA.

Zajonc, R. B. (1965). Social facilitation. *Science*, 149, 269–274.

Zillmann, D. (1991). Television viewing and physiological arousal. In J. Bryant and D. Zillmann (eds.), *Responding to the screen: Reception and reaction processes* (pp. 103–133). Hillsdale, NJ: Lawrence Erlbaum Associates.

➤ Chapter 5

Aristotle. (1984). *The rhetoric and the poetics of Aristotle*. New York: McGraw-Hill.

Dahlbäck, N., Wang, Q. Y., Nass, C., and Alwin, J. (2007). Similarity is more important than expertise: Accent effects in speech interfaces. In *Proceedings of the SIGCHI conference on human factors in computing systems* (pp. 1553-1556). San Jose, CA: ACM.

"Desert Survival Situation" is trademarked by and was used by permission of Human Synergistics.

Fleming, V. (director). (1939). *The wizard of Oz* [Motion picture]. United States: MGM.

Fogg, B. J., and Nass, C. (1997). Do users reciprocate to computers? In *Proceedings of the computer-human interaction (CHI) conference* (pp. 331–332). Atlanta, GA: ACM.

Gilbert, D. T. (1991). How mental systems believe. *American Psychologist*, *46*(2), 107–119.

Goleman, D. (1995). *Emotional intelligence: Why it can matter more than IQ*. New York: Bantam Books.

Hovland, C. I., and Weiss, W. (1951). The influence of source credibility on communication effectiveness. *Public Opinion Quarterly*, *15*(4), 635–650.

Johnson, V. E. (2002). Teacher course evaluations and student grades: An academic tango. *Chance*, *15*(3), 9–16.

Kanter, R. M. (1993). *Men and women of the corporation: New edition* (2nd ed.). New York: Basic Books.

Kaptein, M., Nass, C., and van Bel, D. (unpublished). Better the devil you know than the devil you don't. Stanford University.

Kelley, R. E., and Caplan, J. (1993). How Bell Labs creates star performers. *Harvard Business Review*, July-August, 128–139.

Lafferty, J. C., and Eady, P. M. (1974). *The desert survival problem*. Plymouth, MI: Experimental Learning Methods.

Lee, J. R. (2008). *A threat on the net: Stereotype threat in avatar-represented groups*. Unpublished doctoral dissertation, Stanford University, Stanford, CA.

Leshner, G., Reeves, B., and Nass, C. (1998). Switching channels: The effects of television channels on the mental representation of television. *Journal of Broadcasting & Electronic Media*, *42*(1), 21–33.

Maccoby, E. (1998). *The two sexes: Growing up apart, coming together*. Cambridge, MA: Harvard University Press.

Mahoney, M. J. (1976). *Scientist as subject: The psychological imperative*. Pensacola, FL: Ballinger Publishing.

Morishima, Y., Bennett, C., Nass, C., and Lee, K. M. (unpublished). Effects of (synthetic) voice gender, user gender, and product gender on credibility in e-commerce. Stanford University.

Nass, C., and Brave, S. B. (2005). *Wired for speech: How voice activates and enhances the human-computer relationship*. Cambridge, MA: MIT Press.

Nass, C., Reeves, B., and Leshner, G. (1996). Technology and roles: A tale of two TVs. *Journal of Communication*, *46*(2), 121–128.

Nisbett, R. E., and Wilson, T. D. (1977). The halo effect: Evidence for unconscious alteration of judgments. *Journal of Personality and Social Psychology*, 35(4), 250–256.

Plato. (1992). *Plato: Republic*. Indianapolis, IN: Hackett.

Reeves, B., and Nass, C. (1996). *The media equation: How people treat computers, television, and new media like real people and places*. New York: Cambridge University Press.

Saegert, S., Swap, W., and Zajonc, R. B. (1973). Exposure, context, and interpersonal attraction. *Journal of Personality and Social Psychology*, 25(2), 234–242.

Steele, C. M., and Aronson, J. (1995). Stereotype threat and the intellectual test performance of African Americans. *Journal of Personality and Social Psychology*, 69(5), 797–811.

Index

Page numbers in *italics* refer to illustrations.